Principles of Concurrent Programming

Principles of Concurrent Programming

M. Ben-Ari

Tel-Aviv University

Prentice/Hall International

Englewood Cliffs, New Jersey   London   New Delhi
Singapore   Sydney   Tokyo   Toronto   Wellington

Library of Congress Cataloging in Publication Data

Ben-Ari, M.,   1948–
   Principles of concurrent programming.

   Bibliography: p.
   Includes index.
   1. Parallel processing (Electronic computers)
   I. Title.
   QA76.6.B437   001.64'2   82–3650
   ISBN 0-13-701078-8 (pbk.)   AACR2

British Library Cataloguing in Publication Data

Ben-Ari, M.
   Principles of concurrent programming.
   1. Parallel processing (Electronic computers)
   2. Operating systems (Computers)
   I. Title
   001.64'25   QA76.6
   ISBN 0-13-701078-8

ISBN   0-13-701078-8

PRENTICE-HALL INTERNATIONAL INC., London
PRENTICE-HALL OF AUSTRALIA PTY., LTD., Sydney
PRENTICE-HALL CANADA, INC., Toronto
PRENTICE-HALL OF INDIA PRIVATE LIMITED, New Delhi
PRENTICE-HALL OF JAPAN, INC., Tokyo
PRENTICE-HALL OF SOUTHEAST ASIA PTE., LTD., Singapore
PRENTICE-HALL INC., Englewood Cliffs, New Jersey
WHITEHALL BOOKS LIMITED, Wellington, New Zealand

Printed in the United States of America

10 9 8 7 6

To Margalit

CONTENTS

PREFACE

Concurrent programming—the programming tools and techniques for dealing with parallel processes—has traditionally been a topic in operating systems theory texts. There are several reasons why concurrent programming deserves a book of its own and should be the core of an advanced computer science systems course:

1. Concurrent programming is what distinguishes operating systems and real-time systems from other software systems. Computer science students who have taken courses in programming, data structures, computer architecture and probability can easily master the applications of these disciplines in operating systems, but they need to be introduced to techniques that will enable them to deal with parallelism.

2. I doubt if many of my students will ever design or construct a multiprocessing time-sharing system where algorithms for paging and scheduling are of prime importance. I am certain that they will be designing and constructing real-time systems for mini- and microcomputers. A sound knowledge of concurrent programming will enable them to cope with real-time systems—in particular with the severe reliability requirements that are imposed.

3. There is a trend towards increasing use of abstract concurrency that has nothing to do with the parallelism of actual systems. Data flow diagrams used in software engineering are nothing more than networks of concurrent processes. Traditionally, such a design must be implemented in a sequential language, but UNIX† is a programming system which encourages the use of concurrent processes. Ada‡, the new language designed for the U.S. Department of Defense, includes concurrent programming features as an integral part of the language.

† UNIX is a trademark of Bell Laboratories.
‡ Ada is a trademark of the United States Dept. of Defense.

xiii

4. Finally, concurrent programming is an important topic in computer science research. The basic results in the field are still scattered throughout the literature and deserve to be collected into one volume for a newcomer to the field.

The book requires no prerequisites as such other than computer science maturity (to borrow the term from mathematics). The book is aimed at advanced undergraduate students of computer science. At the Tel Aviv Univeristy, we arrange the course of study so that a student has four semesters of computer science courses which include extensive programming exercises. The material is also appropriate for practicing systems and real-time programmers who are looking for a more formal treatment of the tools of their trade.

I have used the material for half of a weekly four-hour semester course in operating systems—the second half is devoted to the classical subjects: memory management, etc. I should like to see curricula evolve that would devote a quarter or trimester to (theoretical) concurrent programming followed by a project oriented course on operating systems or real-time systems.

The book is not oriented to any particular system or technique. I have tried to give equal time to the most widespread and successful tools for concurrent programming: memory arbiters, semaphores, monitors and rendezvous. Only the most elementary features of Pascal and Ada are used; they should be understandable by anyone with experience in a modern programming language.

Much of the presentation closely follows the original research articles by E. W. Dijkstra and C. A. R. Hoare. In particular, Dijkstra's *Co-operating Sequential Processes* reads like a novel and it was always great fun to lecture the material. This book is an attempt to explain and expand on their work.

Verification of concurrent programs is one of the most exciting areas of research today. Concurrent programs are notorious for the hidden and sophisticated bugs they contain. Formal verification seems indispensable. A novel feature of this book is its attempt to verify concurrent programs rigorously though informally. Hopefully, a student who is introduced early to verification will be prepared to study formal methods at a more advanced level.

One cannot learn any programming technique without practicing it. There are several programming systems that can be used for class exercise. For insititutions where no such system is available, the Appendix contains the description and listing of a simple Pascal-S based concurrent programming system that I have used for class exercise.

ACKNOWLEDGEMENTS

I am grateful to Amiram Yehudai for the many discussions that we have had over the past few years on programming and how to teach it. He was also kind enough to read the entire manuscript and suggest improvements.

I should also like to thank Amir Pnueli for teaching me programming logic and verification, and Henry Hirschberg of Prentice/Hall International for his encouragement during the transition from notes to book.

Also, special thanks to my wife Margalit for drawing the sketches on which the illustrations were based.

Principles of
Concurrent
Programming

1 WHAT IS CONCURRENT PROGRAMMING?

1.1 FROM SEQUENTIAL TO CONCURRENT PROGRAMMING

Figure 1.1 shows an interchange sort program. The program can be compiled into a set of machine language instructions and then executed on a computer. The program is sequential; for any given input (of 40 integers) the computer will always execute the same sequence of machine instructions.

If we suspect that there is a bug in the program then we can debug by tracing (listing the sequence of instructions executed) or by breakpoints and snapshots (suspending the execution of the program to list the values of the variables).

There are better sequential sorting algorithms (see Aho *et al.*, 1974) but we are going to improve the performance of this algorithm by exploiting the possibility of executing portions of the sort in parallel. Suppose that (for $n=10$) the input sequence is: 4, 2, 7, 6, 1, 8, 5, 0, 3, 9. Divide the array into two halves: 4, 2, 7, 6, 1 and 8, 5, 0, 3, 9; get two colleagues to sort the halves simultaneously: 1, 2, 4, 6, 7 and 0, 3, 5, 8, 9; and finally, with a brief inspection of the data, merge the two halves:

0
0, 1
0, 1, 2
. . .

A simple complexity analysis will now show that even without help of colleagues, the parallel algorithm can still be more efficient than the sequential algorithm. In the inner loop of an interchange sort, there are $(n-1) + (n-2) + \ldots + 1 = n(n-1)/2$ comparisons. This is approx. $n^2/2$. To sort $n/2$

```
program sortprogram;
const n=40;
var   a: array[1..n] of integer;
      k: integer;
procedure sort(low,high: integer);
var       i,j, temp: integer;
begin
   for i := low to high−1 do
          for j := i+1 to high do
          if a[j] < a[i] then
          begin
             temp := a[j];
             a[j] := a[i];
             a[i] := temp
          end
end;
begin (* main program *)
   for k := 1 to n do read (a[k]);
   sort (1, n);
   for k := 1 to n do write (a[k])
end.
```

Fig. 1.1.

elements, however, requires only $(n/2)^2/2 = n^2/8$ comparisons. Thus the parallel algorithm can perform the entire sort in twice $n^2/8 = n^2/4$ comparisons to sort the two halves plus another n comparisons to merge. The table in Fig. 1.2 demonstrates the superiority of the new algorithm. The last column shows that additional savings can be achieved if the two sorts are performed simultaneously.

n	$n^2/2$	$(n^2/4)+n$	$(n^2/8)+n$
40	800	440	140
100	5000	2600	1350
1000	500 000	251 000	126 000

Fig. 1.2.

Figure 1.3 is a sequential program for this algorithm. It can be executed on any computer with a Pascal compiler and it can be easily translated into other computer languages.

```
program sortprogram;
const n=20;
      twon=40;
var   a: array[1..twon] of integer;
      k: integer;
procedure sort(low, high: integer);
         (* as before *)
procedure merge(low, middle, high: integer);
var   count1, count2: integer;
      k, index1, index2: integer;
begin
  count1 := low;
  count2 := middle;
  while count1 < middle do
  if a[count1] < a[count2] then
    begin
      write (a[count1]);
      count1 := count1+1;
      if count1 >= middle then
      for index2 := count2 to high do
        write(a[index2])
    end
  else
    begin
      write (a[count2]);
      count2 := count2+1;
      if count2 > high then
        begin
          for index1 := count1 to middle−1 do
            write (a[index1]);
          count1 := middle (* terminate *)
        end
    end
end;
begin (* main program *)
  for k := 1 to twon do read (a[k]);
  sort(1, n);
  sort(n+1, twon);
  merge(1, n+1, twon)
end.
```

Fig. 1.3.

Suppose that the program is to be run on a multiprocessor computer—a computer with more than one CPU. Then we need some notation that can express the fact that the calls *sort*(1,*n*) and *sort*(*n*+1, *twon*) can be executed in parallel. Such a notation is the **cobegin–coend** bracket shown in Fig. 1.4. **cobegin** p_1; . . . ; p_n **coend** means: suspend the execution of the main program; initiate the execution of procedures $p_1, . . . , p_n$ on multiple computers; when all of $p_1, . . . , p_n$ have terminated then resume the main program.

```
program sortprogram;
( *declarations as before *)
begin (* main program *)
  for k := to twon do read(a[k]);
  cobegin
    sort(1, n);
    sort(n+1, twon)
  coend;
    merge(1, n+1, twon)
end.
```

Fig. 1.4.

The programs of Figs. 1.3 and 1.4 are identical except for the **cobegin–coend** in Fig. 1.4. There would be no need for both versions if the definition of **cobegin–coend** was modified. Instead of requiring that the procedures be executed in parallel, **cobegin–coend** becomes a declaration that the procedures *may* be executed in parallel. It is left to the implementation—the system hardware and software—to decide if parallel execution will be done. Processors may be added or removed from the system without affecting the correctness of the program—only the time that it would take to execute the program.

The word *concurrent* is used to describe processes that have the potential for parallel execution. We have shown how an algorithm can be improved by identifying procedures that may be executed concurrently. While the greatest improvement is obtained only under true parallel execution, it is possible to ignore this implementation detail without affecting the superiority of the concurrent algorithm over the sequential algorithm.

1.2 CONCURRENT PROGRAMMING

Concurrent programming is the name given to programming notations and techniques for expressing potential parallelism and for solving the resulting synchronization and communication problems. Implementation of parallelism is a topic in computer systems (hardware and software) that is essentially independent of concurrent programming. Concurrent programming is important because it provides an abstract setting in which to study parallelism without getting bogged down in the implementation details. This abstraction has proved to be so useful in writing clear, correct software that modern programming languages offer facilities for concurrent programming.

The basic problem in writing a concurrent program is to identify which activities may be done concurrently. If the *merge* procedure is also included in the **cobegin–coend** bracket (Fig. 1.5), the program is no longer correct. If you merge the data in parallel with sorting done by your two colleagues, the scenario of Fig. 1.6 might occur.

```
cobegin
    sort(1, n);
    sort(n+1, twon);
    merge(1, n+1, twon)
coend
```

Fig. 1.5.

	Colleague1	Colleague2	You
Initially	4, 2, 7, 6, 1	8, 5, 0, 3, 9	–
Colleague1 exchanges	2, 4, 7, 6, 1	8, 5, 0, 3, 9	–
Colleague2 exchanges	2, 4, 7, 6, 1	5, 8, 0, 3, 9	–
You merge	,,	,,	2
You merge	,,	,,	2, 4
You merge	,,	,,	2, 4, 5

Fig. 1.6.

However, *merge* could be a concurrent process if there were some way of synchronizing its execution with the execution of the sort processes (Fig. 1.7).

```
while count1 < middle do
    wait until i of procedure call sort(1,n) is greater than count1 and i of
    procedure call sort(n+1, twon) is greater than count2 and only then:
    if a[count1] < a[count2] then
        . . .
```

Fig. 1.7.

Parallelism is important not only for the improvement that can be achieved in program performance but also in program quality. Consider the following problem:

> Read 80-column cards and print them on 125-character lines. However, every run of $n = 1$ to 9 blank spaces is to be replaced by a single blank followed by the numeral n.

This program is difficult to write as a sequential program. There are many interacting special cases: a run of blanks overlapping the end of a card, the pair blank-n overlapping the end of a line and so on. One way to improve the clarity of the program would be to write three separate programs: one to read cards and write a stream of characters onto a temporary file; a second program to read this character stream and modify runs of blanks, writing the new stream onto a second temporary file; and a third program to read the second temporary file and print lines of 125 characters each.

This solution is not acceptable because of the high overhead of the temporary files. However, if the three programs could be run concurrently (not necessarily in parallel) and communications paths could be established between them, then the programs would be both efficient and elegant.

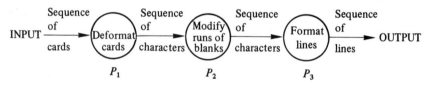

Processes P_1, P_2, P_3 may execute concurrently

1.3 CORRECTNESS OF CONCURRENT PROGRAMS

Concurrent programming is much more difficult than sequential programming because of the difficulty of ensuring that a concurrent program is correct. Consider the sequential sort programs of Figs. 1.1 and 1.3: if they were tested on several sets of input data then we would feel confident that they are correct. Guidelines for testing would be to include a sorted input, a reversed input, an input all of whose elements are identical and so on. A run-of-the-mill bug (such as an incorrect **for**-loop limit) would seldom escape detection.

The scenario in Fig. 1.6. illustrates that the concurrent program of Fig. 1.5 is incorrect. However, this program is not a sequential program and other scenarios exist. If the processors assigned to sort are sufficiently rapid then *merge* may always be working on sorted data. In that case, no amount of testing would detect any problem. One day (perhaps months after the program has been put into production) an improved component in the computer system causes the merge to speed up and then the program gives incorrect answers as demonstrated in 1.6. Of course the natural reaction is:

"This program worked yesterday so the new component must be at fault."

A *scenario* is a description of a possible execution sequence of a program and shows how a computer might "act out" a program. It is usually used to show that a program is incorrect: since the computer may execute the program in a manner that produces the wrong answer, the program cannot be correct.

Conversely, how can we show that the concurrent program in Fig. 1.4 is correct? It no longer makes sense to look for and test paths that can be execution sequences. At times, there may be two such sequences caused by parallel execution of the algorithm.

Sequential programming has a well-developed proof theory. Assertions are made about the state of the computer (i.e. the values of the variables and the program counter) before and after executing an instruction, and these are then combined into a logical proof. In concurrent programming, this method needs to be modified because the programs can interfere with each other.

The correctness assertions for procedures *sort* and *merge* of the previous sections are elementary to state and prove:

> *sort* input assertion: *a* is an array of integers,
> *sort* output assertion: *a* is "sorted", i.e. *a* now contains a permutation of the original elements and they are in ascending order,
> *merge* input assertion: the two halves of *a* are "sorted",
> *merge* output assertion: the elements of *a* have been written in ascending order.

The correctness of the program in Fig. 1.1. is immediate from the correctness of procedure *sort*. The correctness of 1.3 is easily obtained by *concatenating* the correctness proofs of *sort* and *merge*. The correctness of Fig. 1.4 needs a new technique. We have to be able to express the fact that the two instances of *sort* do not interfere with one another. The program in Fig. 1.5 is incorrect though the procedures comprising it are correct; unfortunately, they interact in a manner which makes the program incorrect. The program in 1.7 is correct but new ideas are needed to be able to reason about synchronization.

1.4 INTERLEAVING

Interleaving is a logical device that makes it possible to analyze the correctness of concurrent programs. Suppose that a concurrent program P consists of two processes P_1 and P_2. Then we say that P executes any one of the execution sequences that can be obtained by interleaving the execution sequences of the two processes. It is as if some supernatural being were to execute the instructions one at a time, each time flipping a coin to decide whether the next instruction will be from P_1 or P_2.

We claim that these execution sequences exhaust the possible behaviors of P. Consider any instructions I_1 and I_2 from P_1 and P_2, respectively. If I_1 and I_2 do not access the same memory cell or register then it certainly does not matter if I_1 is executed before I_2, after I_2 or even simultaneously with I_2 (if the hardware so allows). Suppose on the other hand that I_1 is "Store 1 into memory cell M" and that I_2 is "Store 2 into memory cell M". If I_1 and I_2 are executed simultaneously then the only reasonable assumption is that the result is consistent. That is, cell M will contain either 1 or 2 and the computer does not store another value (such as 3) of its own volition.

If this were not true then it would be impossible to reason about concurrent programs. The result of an individual instruction on any given data cannot depend upon the circumstances of its execution. Only the external behavior of the system may change—depending upon the interaction of the instructions through the common data. In fact, computer hardware is built so that the result of executing an individual instruction is consistent in the way just defined.

Thus, if the result of the simultaneous execution of I_1 and I_2 is 1 then this is the same as saying that I_1 occurred before I_2 in an interleaving and conversely if the result is 2.

Interleaving does not make the analysis of concurrent programs simple. The number of possible execution sequences can be astronomical. Nevertheless, interleaved execution sequences are amenable to formal methods and will allow us to demonstrate the correctness of concurrent programs.

1.5   THE ORIGIN OF OPERATING SYSTEMS

Concurrent programming, though generally applicable, grew out of problems associated with operating systems. This section outlines the development of such systems so that the background to the growth of concurrent programming can be appreciated.

It is not often that an obsolete technology reappears. Programmable pocket calculators have resurrected machine language programming: absolute addresses must be used for data and labels. On the other hand, the owner is not constrained to working during the hours that the computer center is open.

While the pocket calculator is a marvel of electronics, machine language programming directly on the computer is slow and difficult. In the 1950s, when computers were few and expensive, there was great concern over the waste caused by this method. If you signed up to sit at the computer console from 0130 to 0200 and you spent 25 minutes looking for a bug, this 25 minutes of computer idle time could not be recovered. Nor was your colleague who signed up for 0200–0230 likely to let you start another run at 0158.

If we analyze what is happening in the terms of the previous sections we see that the manual procedures that must be performed—mounting tapes, setting up card decks, or changing places at the console—are disjoint from the actual computation and can be performed concurrently with the computer's processing.

The second generation of computers used a supervisor program to batch jobs. A professional computer operator sat at the console. Programmers prepared card decks which were concatenated into "batches" that were fed into the computer once an hour or so. The increase in throughput (a measure of the efficiency of a computer; it is the number of jobs—suitably weighted—that can be run in a given time period) was enormous—the jobs were run one after another with no lost minutes. The programmers, however, lost the ability to dynamically track the progress of their programs since they no longer sat at the computer console. In the event of an error in one job, the computer simply commenced execution of the next job in the batch, leaving the programmer to puzzle out what happened from core dumps. With a turnaround time (the amount of time that elapses between a job being submitted for execution and the results being printed) of hours or days, the task of programming became more difficult even though certain aspects were improved by high-level languages and program libraries.

Despite this improvement in throughput, systems designers had noticed another source of inefficiency not apparent to the human eye. Suppose that a computer can execute one million instructions per second and that it is connected to a card reader which can read 300 cards per minute (= one card in 1/5 second). Then from the time the *read* instruction is issued until the time the card has been read, 200 000 instructions could have been executed. A program to read a deck of cards and print the average of the numbers punched in the cards will spend over 99% of its time doing nothing even though 5 cards per second seems very fast.

The first solution to this problem was *spooling*. The I/O speed of a magnetic tape is much greater than that of the card reader and the line printer that are the interface between the computer and the programmer. We can decompose the operation of the computer into three processes: a process to read cards to tape; a process to execute the programs on the tape and write the results onto a second tape; and a process to print the information from the second tape. Since these processes are disjoint (except for the exchange of the tapes after processing a batch), the throughput can be greatly increased by running each process on a separate computer. Since very simple computers can be used to transfer information to and from the magnetic tape, the increase in cost is not very great compared to the savings achieved by more efficient use of the main computer.

Later generations of computer systems have attacked these problems by switching the computer among several computations whose programs and data are held simultaneously in memory. This is known as *multiprogramming*. While I/O is in progress for program P_1 the computer will execute several thousand instructions of program P_2 and then return to process the data obtained for P_1. Similarly, while one programmer sitting at the terminal of a time-sharing system† is thinking, the computer will switch itself to execute the program requested by a second programmer. In fact, modern computer systems are so powerful that they can switch themselves among dozens or even hundreds of I/O devices and terminals. Even a minicomputer can deal with a dozen terminals.

The importance of the concept of interleaved computations mentioned in the previous section has its roots in these multiprogrammed systems. Rather than attempt to deal with the global behavior of the switched computer, we will consider the actual processor to be merely a means of interleaving the computations of several processors. Even though multiprocessor systems—systems with more than one computer working simultaneously—are becoming more common, the interleaved computation model is still appropriate.

The sophisticated software systems that are responsible for multiprogramming are called *operating systems*. The term operating system is often used to cover all manufacturer-provided software such as I/O programs and compilers and not just the software responsible for the multiprogramming.

While the original concern of operating system designers was to improve throughput, it soon turned out that the throughput was affected by numerous system "crashes" when the system stopped functioning as it was supposed to and extensive recovery and restart measures delayed execution of jobs. These defects in the operating systems were caused by our inadequate understanding of how to execute several programs simultaneously and new design and programming techniques are needed to prevent them.

1.6   OPERATING SYSTEMS AND CONCURRENT PROGRAMMING

If you could sense the operation of a computer that is switching itself every few milliseconds among dozens of tasks you would certainly agree that the computer seems to be performing these tasks simultaneously even though we know that the computer is interleaving the computations of the various tasks. I now argue that it is more than a useful fiction to assume that the computer is in fact performing its tasks concurrently. To see why this is so, let us consider task switching in greater detail. Most computers use interrupts

---

† A time-sharing system is a computer system that allows many programmers to work simultaneously at terminals. Each programmer may work under the illusion that the computer is working for him alone (though the computer may seem to be working slowly if too many terminals are connected).

for this purpose. A typical scenario for task switch by interrupts is as follows. Program P_1 makes a read request and then has its execution suspended. The CPU may now execute program P_2. When the read requested by P_1 has been completed, the I/O device will interrupt the execution of P_2 to allow the operating system to record the completion of the read. Now the execution of either P_1 or P_2 may be resumed.

The interrupts occur asynchronously during the execution of programs by the CPU. By this is meant that there is no way of predicting or coordinating the occurence of the interrupt with the execution of any arbitrary instruction by the CPU. For example, if the operator who is mounting a magnetic tape happens to sneeze, it may delay the "tape ready" signal by 8.254387 seconds. However, if he is "slow" with his handkerchief, the delay might be 8.254709 seconds. Insignificant as that difference may seem, it is sufficient for the CPU to execute dozens of instructions. Thus for all practical purposes it makes no sense to ask: "What is the program that the computer is executing?" The computer is executing any one of a vast number of execution sequences that may be obtained by arbitrarily interleaving the execution of the instructions of a number of computer programs and I/O device handlers.

This reasoning justifies the abstraction that an operating system consists of many processes executing concurrently. The use of the term process rather than program emphasizes the fact that we need not differentiate between ordinary programs and external devices such as terminals. They are all independent processes that may, however, need to communicate with each other.

The abstraction will try to ignore as many details of the actual application as possible. For example, we will study the producer–consumer problem which is an abstraction both of a program producing data for consumption by a printer and of a card reader producing data for consumption by a program. The synchronization and communication requirements are the same for both problems even though the details of programming an input routine are rather different from the details of an output routine. Even as new I/O devices are invented, the input and output routines can be designed within the framework of the general producer–consumer problem.

On the other hand, we assume that each process is a sequential process. It is always possible to refine the description of a system until it is given in terms of sequential processes.

The concurrent programming paradigm is applicable to a wide range of systems, not just to the large multiprogramming operating systems that gave rise to this viewpoint. Moreover, every computer (except perhaps a calculator or the simplest microcomputer) is executing progams that can be considered to be interleaved concurrent processes. Minicomputers are supplied with small multiprogramming systems. If not, they may embedded in real-time systems‡ where they are expected to concurrently absorb and process dozens of different asynchronous external signals and operator commands. Finally, networks of interconnected computers are becoming common. In this case true parallel processing is occurring. Another term used is *distributed processing* to emphasize that the connected computers may be physically separated. While the abstract concurrency that models switched systems is now well understood, the behavior of distributed systems is an area of current research.

1.7 AN OVERVIEW OF THE BOOK

Within the overall context of writing correct software this book treats the single, but extremely important, technical point of synchronization and communication in concurrent programming. The problems are very subtle; ignoring the details can give rise to spectacular bugs. In Chapter 2 we shall define the concurrent programming abstraction and the arguments that justify each point in the definition. The abstraction is sufficiently general that it can be applied without difficulty to real systems. On the other hand it is sufficiently simple to allow a precise specification of both good and bad behavior of these programs.

Formal logics exist which can formulate specifications and prove properties of concurrent programs in this abstraction though we will limit ourselves to informal or at most semi-formal discussions. The fact that the discussion is informal must not be construed as meaning that the discussion is imprecise. A mathematical argument is considered to be precise even if it is not formalized in logic and set theory.

The basic concurrent programming problem is that of mutual exclusion. Several processes compete for the use of a certain resource such as a tape drive but the nature of the resource requires that only one process at a time actually accessed the resource. In other words, the use of the resource by one process excludes other processes from using the resource. Chapter 3 presents a series of attempts to solve this problem culminating in the solution known as Dekker's algorithm. The unsuccessful attempts will each point out a possible "bad" behavior of a concurrent program and will highlight the differences between concurrent and sequential programs.

Dekker's algorithm is itself too complex to serve as a model for more complex programs. Instead, synchronization primitives are introduced. Just as a disk file can be copied onto tape by a single control language command

---

‡ Wheareas a time-sharing system gives the user the ability to use all the resources of a computer, the term real-time system is usually restricted to systems that are required to respond to specific pre-defined requests from a user or an external sensor. Examples would be air-traffic control systems and hospital monitoring systems.

or a file can be read by writing *read* in a high level language, so we can define programming language constructs for synchronization by their semantic definition—what they are supposed to do—and not by their implementation. We shall indicate in general terms how these primitives can be implemented but the details vary so much from system to system that to fully describe them would defeat our purpose of studying an abstraction. Hopefully, it should be possible for a "casual" systems programmer to write concurrent programs without knowing how the primitives are implemented. A model implementation is described in the Appendix.

Chapter 4 commences the study of high level primitives with E. W. Dijkstra's semaphore. The semaphore has proved extraordinarily successful as the basic synchronization primitive in terms of which all others can be defined. The semaphore has become the standard of comparison. It is sufficiently powerful that interesting problems have elegant solutions by semaphores and it is sufficiently elementary that it can be successfully studied by formal methods. The chapter is based on the producer–consumer problem mentioned above; the mutual exclusion problem can be trivially solved by semaphores.

Most operating systems have been based on monolithic monitors. A central executive, supervisor or kernel program is given sole authority over synchronization. Monitors, a generalization of this concept formalized by Hoare, are the subject of Chapter 5. The monitor is a powerful conceptual notion that aids in the development of well structured, reliable programs. The problem studied in this chapter is the problem of the readers and the writers. This is a variant of the mutual exclusion problem in which there are two classes of processes: writers which need exclusive access to a resource and readers which need not exclude one another (though as a class they must exclude all writers).

The advent of distributed systems has posed new problems for concurrent programming. C. A. R. Hoare has proposed a method of synchronization by communication (also known as synchronization by rendezvous) appropriate for this type of system. The designers of the Ada programming language have chosen to incorporate in the language a variant of Hoare's system. Anticipating the future importance of the Ada language, Chapter 6 studies the Ada rendezvous.

A classic problem in concurrent programming is that of the Dining Philosophers. Though the problem is of greater entertainment value than practical value, it is sufficiently difficult to afford a vehicle for the comparison of synchronization primitives and a standing challenge to proposers of new systems. Chapter 7 reviews the various primitives studied by examining solutions to the problem of the Dining Philosophers.

Programming cannot be learned without practice and concurrent programming is no exception. If you are fortunate enough to have easy access to a minicomputer or to a sophisticated simulation program, there may be no difficulty in practicing these new concepts. If not, the Appendix describes in full detail an extremely simple simulator of concurrency that can be used for class exercise. In any case, the Appendix can serve as an introduction to implementation of concurrency.

The book ends with an annotated bibliography suggesting further study of concurrent programming.

1.8 PROGRAM NOTATION

The examples in the text will be written in a restricted subset of Pascal-S, which is itself a highly restricted subset of Pascal. This subset must of course be augmented by constructs for concurrent programming. It is intended that the examples be legible to any programmer with experience in Pascal, Ada, C, Algol, or PL/I.

The implementation kit in the Appendix describes an interpreter for this language that will execute the examples and that can be used to program the exercises. The language in the kit contains more Pascal language features than are used in the text of the book and thus users of the kit are assumed to be able to program in sequential Pascal. These extra features are necessary in order to use the kit to solve the exercises, although the exercises themselves could be programmed in other languages that provide facilities for concurrent programming.

The examples in the chapter on monitors are standard and can be adapted to the many systems that provide the monitor facility such as Concurrent Pascal, Pascal-Plus, or CSP/k. The examples in the chapter on the Ada rendezvous are executable in Ada.

We now present a sketch of the language that should be sufficient to enable programmers unfamiliar with Pascal to understand the examples.

1. Comments are inserted between (* and *).

2. The first line in a program should be

   program *name*;

3. Symbolic names for constants may be declared by the word **const** followed by the constant itself:

   const *twon*=40;

4. All variables in each procedure in the main program must be declared by the word **var** followed by the names of the variables and a type:

   var *i, j, temp*: *integer*;
       found: *boolean*;

The available types are : *integer, boolean* (with constants *true* and *false*) and arrays:

> **var** *a*:**array**[*lowindex...highindex*] **of** *integer*;

5. Following the declaration of the variables, procedures and functions may be declared: **procedure** *name* (*formal parameters*); and **function** *name* (*formal parameters*): *returntype*; . The formal parameter definition has the same form as that of a variable list:

> **procedure** *sort* (*low,high*: *integer*);
> **function** *last*(*index*: *integer*): *boolean*;

6. The body of the main program or of procedures is a sequence of statements separated by semi-colons between **begin** and **end**. The main program body is terminated by a period and the procedure bodies by semi-colons. The usual rules on nested scopes apply.

7. The statements are:

> *assignment statement*
> **if** *boolean-expression* **then** *statement*
> **if** *boolean-expression* **then** *statement* **else** *statement*
> **for** *index-variable* := *lowindex* **to** *highindex* **do** *statement*
> **while** *boolean-expression* **do** *statement*
> **repeat** *sequence-of-statements* **until** *boolean-expression*

The syntactic difference between **while** and **repeat** is that **while** takes a single statement and **repeat** takes a sequence of statements (separated by semi-colons). The semantic difference is that the **while** tests before the loop is done and **repeat** tests afterwards. Thus **repeat** executes its loop at least once.

8. A sequence of statements may be substituted for "statement" in the above forms by enclosing the sequence of statements in the bracket **begin** ... **end** to form a single "compound" statement:

> **if** *a*[*j*] < *a*[*i*] **then**
>   **begin**
>   *temp* := *a*[*j*];
>   *a*[*j*] := *a*[*i*];
>   *a*[*i*] := *temp*
>   **end**

In detail this is read: **if** the boolean expression (*a*[*j*] < *a*[*i*]) has the value *true*, **then** execute the compound statement which is a sequence of three assignment statements. If the expression is *false*, then the (compound) statement is not executed and the execution continues with the next statement.

9. Assignment statements are written variable := expression. The variable may be a simple variable or an element of an array: $a[i]$. The type (*integer* or *boolean*) of the expression must match that of the variable. Integer expressions are composed of integer variables and constants using the operators: $+$, $-$, $*$, **div** (integer divide with truncation) and **mod**. Boolean expressions may be formed from relations between integer expressions: $=$, $<>$ (not equal), $<$, $>$, $<=$ (less than or equal) $>=$ (greater than or equal). The boolean operators **and**, **or** and **not** may be used to form compound boolean expressions.

10. For those who know Pascal we list here the additional features that are defined in the language of the implementation kit some of which will be necessary if you plan to write any programs using the kit.

   (a) Type declarations. Since there are no scalar, subrange or record types, this is mostly useful for array types:
   type *sortarray* = **array**[1..*n*] **of** *integer*
   var *a*: *sortarray*;
   (b) Character constants and variables of type **char**.
   (c) Multidimensional arrays (arrays of arrays).
   (d) A parameter may be passed by reference rather than value by prefixing the formal parameter by **var**.
   (e) Recursive functions and procedures.
   (f) I/O may be performed only on the standard textfiles *input* and *output*. To ensure that you do not forget this restriction, the declaration of external files in the **program** card has been removed. *read*, *readln*, *write*, *writeln*, *eoln*, *eof* (all without a file parameter) function as in Pascal. Only the default field widths may be used in a *write*, which will, however, accept a 'string' as a field to be printed:
   writeln ('the answer is', *n*).

1.9   EXERCISES†

1.1   Write a two-process concurrent program to find the mean of *n* numbers.

1.2   Write a three-process concurrent program to multiply 3×3 matrices.

1.3   Each process of the matrix multiply program executes three multiplications and two additions for each of three rows or altogether 15 instructions. How many execution sequences of the concurrent program may be obtained by interleaving the executions of the three processes?

† Slightly harder exercises are marked throughout the book with an asterisk (*).

1.4 Perform a similar analysis for *sortprogram*. You will have to make some assumptions on the number of interchanges that will be done.

1.5 Test the concurrent *sortprogram* of Fig. 1.3.

1.6 Test the concurrent *sortprogram* of Fig. 1.4 which has the *merge* defined as a third process. Run the program several times with exactly the same data.

1.7 Run the program in Fig. 1.8 several times. Can you explain the results?

```
program increment;
const m = 20;
var n: integer;
procedure incr;
var i: integer;
begin
  for i := 1 to m do n := n+1
end;
begin (* main program *)
  n := 0;
  cobegin
    incr; incr
  coend;
  writeln (' the sum is ', n)
end.
```

Fig. 1.8.

2 THE CONCURRENT PROGRAMMING ABSTRACTION

2.1 INTRODUCTION

Concurrent programming is not the study of operating systems or real-time systems, but of abstract programming problems posed under certain rules. Concurrent programming was motivated by the problems of constructing operating systems, and its examples are abstract versions of such problems. Most importantly, the rules of concurrent programming are satisfied in many systems and thus its techniques can be used in real systems. Components of a system which are not amenable to concurrent programming techniques should be singled out for extremely careful design and implementation.

Chapter 1 gave the definition of a concurrent program. It consists of several sequential processes whose execution sequences are interleaved. The sequential programs are not totally independent – if they were so there would be nothing to study. They must communicate with each other in order to synchronize or to exchange data.

The first means of communication that we shall study is the *common memory*. This is appropriate for the pseudo-parallel switched computers where all processes are running on the same processor and using the same physical memory. It is also used on some truly parallel systems such as the CDC Cyber computers where even though one CPU and ten PP's (peripheral processors) are simultaneously executing separate programs, synchronization is accomplished by having the PP's read and write the CPU's memory. In our abstraction, common memory will be represented simply by global variables accessible to all processes.

Common memory can also be used to hold access-restricted procedures. Access to these procedures is, in effect, allocated to a process. This is the way most third generation operating systems were implemented. The "system" programs can only be called by special instructions which ensure that only one process at a time is executing a system program.

With the introduction of distributed computing it is no longer valid to assume that a common central memory exists. Chapter 5 discusses concurrent programming by means of sending and receiving signals instead of reading and writing a common variable or executing a common procedure. Synchronization by message-passing has been used on several experimental systems for single-processor computers but this approach has not been widely accepted because of the possible inefficiency of message-passing compared with simpler systems. Of course, distributed systems have no choice.

2.2 MUTUAL EXCLUSION

Mutual exclusion is one of the two most important problems in concurrent programming because it is the abstraction of many synchronization problems. We say that activity A_1 of process P_1 and activity A_2 of process P_2 must exclude each other if the execution of A_1 may not overlap the execution of A_2. If P_1 and P_2 simultaneously attempt to execute their respective activities, A_i, then we must ensure that only one of them succeeds. The losing process must block; that is, it must not proceed until the winning process completes the execution of its activity A.

The most common example of the need for mutual exclusion in real systems is resource allocation. Obviously, two tapes cannot be mounted simultaneously on the same tape drive. Some provision must be made for deciding which process will be allocated a free drive and some provision must be made to block processes which request a drive when none is free. There is an obvious solution: run only one job at a time. But this defeats one of the main aims of concurrent programming – parallel execution of several processes.

Meaningful concurrency is possible only if the processes are loosely connected. The loose connection will manifest itself by the need for short and occasional communication. The abstract mutual exclusion problem will be expressed:

> *remainder*
> *pre-protocol*
> *critical section*
> *post-protocol.*

"*remainder*" will be assumed to represent some significant processing. Occasionally, i.e. after the completion of *remainder*, the process needs to enter a short *critical section*. It will execute certain sequences of instructions, called *protocols* before and possibly after the critical section. These protocols will ensure that the critical section is in fact executed so as to exclude all other critical secitons. Of course, just as the critical section should be short relative to the main program in order to benefit from concurrency, the protocols must also be relatively short. The protocols represent the overhead paid for concurrency. Hopefully, if the critical sections and the protocols are sufficiently short then the significant processing abstracted as *remainder* can be overlapped thus justifying the design of the multiprogramming system.

There is another, more important, reason for requiring loose connection among concurrent processes and that is to ensure reliability. We want to be assured that if there is a bug in one of the processes, then it will not propagate itself into a system "crash". It should also be possible to gracefully degrade the performance of a system if an isolated device should fail ("fail-soft"). It would be absurd to have a system crash just because one tape drive became faulty.

The abstract requirement will be that, if a process abnormally terminates outside the critical section then no other process should be affected. (For this purpose the protocols are considered to be part of the critical section.) Since the critical section is where the communication is taking place, it is not reasonable to require the same of the critical sections. We might use the following metaphor. If a runner in a relay race fell after he has passed the baton then the race should not be affected. It is unreasonable to hope that the race is unaffected if the fall occurred at the critical moments during the baton exchange.

This restriction is not unreasonable even in practice. Critical sections such as disk I/O will often be executed by common system routines or by compiler–supplied routines which have been written by a competent systems programmer. The probability of a software error in such a routine should be much smaller than in a run-of-the-mill program.

2.3 CORRECTNESS

What does it mean for concurrent programs to be correct? An ordinary program is correct if it halts and prints the "right" answer. In general, you will know a "right" answer if you see one. This is also true of some concurrent programs such as *sortprogram*.

On the other hand, the single most distinguishing feature of an operating system or real-time system is that it must never halt. The only way to halt a typical operating system is to push the start button on the computer panel. An operating system prints nothing of its own (except some non-essential logging and accounting data). Thus when studying operating systems, we must revamp our notions of what it means for a program to be correct. Since most concurrent programming is done within operating systems and real-time systems, this is the area to be studied in this book.

An obvious truism is that any program must do what its specifications say that it is supposed to do. This is no less true in the case of concurrent programs though the specifications are radically different from those of sequential programs. In our abstraction, we shall distinguish two types of correctness properties: *safety* properties and *liveness* properties.

Safety properties are those required by the static portions of the specifications. These are often the only requirements explicitly specified. Mutual exclusion is a safety property: the requirement that critical sections exclude one another is absolute and does not change during the execution of the program. The safety property of the producer–consumer problem is that the consumer must consume every piece of data produced by the producer and that it must do so in the order in which they were produced.

Safety properties are akin to what is known in the theory of sequential programs as *partial correctness*: if the program terminates, the answers must be "correct". Safety properties are usually relatively easy to show. They are explicitly required by the specifications and programs are designed to meet these specifications. You can always achieve more safety by giving up some concurrency and letting more segments of the process execute sequentially.

Violation of mutual exclusion is the cause of most operating system crashes. Dynamic memory allocation is frequently involved: a process may be convinced that it knows the location of a certain table in memory while in fact the table has been removed and its memory allocated to another process.

At a computer center known to the Author, an unscrupulous programmer managed to insert into the system a program that, on command, would give his programs the highest priority. However, he did not know that mutual exclusion protocols were required because the scheduling table was not fixed in memory. Thus the execution of his command would often write into some other table of the system causing a crash. Since the crash occurred many minutes later, the memory dumps gave no clue as to what had happened. After several weeks, the problem was solved by noting the strange command on the system log. If such an error had been made in a regular systems program it would have been even more difficult to catch.

Liveness, on the other hand, deals with dynamic properties. It is akin to sequential programming's *total correctness*: the program terminates and the answer is "correct". In concurrent systems, liveness means that if something is supposed to happen then eventually it will happen. If a process wishes to enter its critical section then eventually it will do so. If a producer produces data then eventually the consumer will consume it.

The most serious breach of liveness is the global form known as *deadlock*.† Deadlock means that the computer is no longer doing any (useful) work. A loop that searches for a free tape drive when none is ever going to be available is not useful work. If all processes are suspended or in such loops, then the computer is said to be deadlocked. This system hang is catastrophic since all users in a multiprogrammed system are affected.

The following deadlock scenario actually occurred in an operating system. The system had multiple processors. Each job wrote its accounting data into a common area of memory which was written to a disk file when it filled. The write procedure ran on whatever processor was free. If all the processors tried to write accounting data at the same time, there was no free processor to write to disk. Since no program voluntarily relinquished its processor, the system deadlocked. A typical solution in this case is to use the computer console terminal to fool a process into thinking that the memory area is empty. Of course, some data will be lost but this is usually preferable to a deadlocked system and restarting all currently running programs.

A local breach of liveness is called *lockout* or (individual) *starvation*. In this case there is always some process which can progress but some identifiable process is being indefinitely delayed. Lockout is less serious than deadlock since the computer is still doing some (presumably) useful work. On the other hand, lockout is difficult to discover and correct because it can happen only in complex scenarios where some processes unwittingly "conspire" to deny a resource to a hapless process.

For reasons to be discussed in the next section, we limit our discussion to qualitative liveness which means that the word "eventually", used in the definition of liveness, means exactly that: within an unspecified but finite length of time.

Between the complete disregard of time by the liveness concept and the introduction of explicit time, is the concept of *fairness*: a process wishing to progress must get a fair deal relative to all other processes. Fairness is more difficult to define precisely and we will mention it only occasionally. In addition, many systems will consciously be "unfair" by designing a priority scheme to favor some processes over others.

2.4   TIMING

We make no assumptions concerning the absolute or relative speeds at which the processes are executed. This statement seems rather shocking at first. However, failure to follow this restriction in systems design can cause

---

† Deadlock is theoretically considered to be a safety property because it is something that should never happen. However, absence of deadlock can be shown by proving that there is always at least one live process.

serious bugs. The statement is shocking because of the naive preoccupation with efficiency that most programmers have. After all, we know that a disk is slower than a CPU and it is tempting to take advantage of that fact in designing not only the structure of the system but also the synchronization details.

One reason for ignoring timing is that our intuition is not able to cope with the scale involved. There are only about $\frac{1}{2}$ million minutes in a year but there are a million microseconds in a second. It is folly to state that: process P_1 ought to be able to finish its critical section before process P_2 finishes its non-critical section, unless such a statement is backed up by a detailed calculation.

A second reason for ignoring timing is that time-dependent bugs are extremely difficult to identify and correct. Qualitative treatments as presented in this book, are insufficient; timing calculations must be included. Each such bug can mean weeks of work for a team of programmers. We shall pay any reasonable penalty in efficiency to obtain a reliable system. You have to save an enormous number of microseconds to make a half-hour system crash worthwhile.

Finally there is an important practical reason to design a software system that is independent of timing assumptions and that is the dynamic nature of a computer configuration. Even if able to carry out the calculations necessary to ensure reliability in a time-dependent system, one is at the mercy of any configuration change. Theoretically, the addition of a single terminal could invalidate the reliability of the system. This is not too far-fetched. A computer manufacturer once started to market a modernized component that ran at twice the speed of the original component. The first customers received a notice that in effect said: "Thanks for buying our double-speed component but please run it at the original speed for a few months while we comb the operating system for time dependencies"! A system may certainly require some time-dependent processing but these functions should be clearly identified and isolated to ensure the flexibility and realiability of the system.

The advantages that accrue to a time-independent (asynchronous) system can be demonstrated on the hardware level by the architecture of the DEC PDP-11 computers. Instead of synchronizing access to memory and I/O devices by a clock, all data in the basic PDP-11 is transferred asynchronously according to a protocol that is not too different conceptually from what we will be doing. The overhead of the protocol means that for any given electronic technology, the PDP-11 will be somewhat slower than a computer built to a synchronous design. On the other hand, the architecture has proved to be extremely flexible; for example, memory modules of arbitrary speeds can be freely mixed. If a new technology produces memory 1.267 faster than before, such a module can be added to a current system and will respond to the protocol just that much faster. A synchronous system can generally mix only modules whose speeds are multiples of each other.

In the abstraction, certain assumptions will be made to avoid meaningless pathologies. Since infinite delay is indistinguishable from a halt, we will assume (globally) that, if there is at least one process ready to run, then some (unspecified) process is allowed to run within a finite time. We also assume (locally) that if a process is allowed to run in its critical section then it will complete the execution of the critical section in a finite period of time.

On the other hand, we allow ourselves to use the adversary approach in checking for possible bugs. A concurrent program suffers from deadlock if it is possible to devise a scenario for deadlock under the sole finiteness assumptions of the previous paragraph. If someone offers you a concurrent program, you can tailor your counterscenario specifically to the given program; you are an "adversary" allowed to plot against the program.

2.5   IMPLEMENTING PRIMITIVE INSTRUCTIONS

Our solutions to the mutual exclusion problem will always cheat by making use of mutual exclusion provided on a lower level—the hardware level. Just as the user of a high level language need not know how a compiler works as long as he is provided with an accurate description of the syntax and semantics of the language, so we will not concern ourselves with how the hardware is implemented as long as we are supplied with an accurate description of the syntax and semantics of the architecture. Presumably the same thing happens at lower levels—the computer logic designer need not know exactly how an integrated circuit is implemented; the integrated circuit designer need only concern himself with the electronic properties of semiconductors and need not know all the details of the quantum physics that explain these properties.

In common memory systems there is an *arbiter* which provides for mutual exclusion in the access to an individual memory word. The word "access" is a generic term for read and write or, as they are usually called, Load and Store corresponding to the assembler instructions for these actions. The arbiter ensures that in case of overlap among accesses, mutual exclusion is obtained by executing the accesses one after the other. The order of the accesses is not guaranteed to the programmer. On the other hand, the consistency of the access is ensured as described in Chapter 1.

Note that the access to a single word is an action that may not be apparent in a high level language. Suppose that n is a global variable that is initially zero and is used as a counter by several processes executing the instruction: $n := n+1$. The compiler compiles such a statement into the three assembler instructions:

    Load n
    Add   1
    Store n

Consider now the following scenario. The value of n is 6. P_1 executes Load n and then P_2 also executes Load n. P_1 increments the value of n in its internal register to obtain 7. Similarly, P_2 obtains the value 7 in its internal register. Finally, the two processes execute the Store instruction in succession and the value 7 is stored twice. Hence the final value of n is 7. That is we have incremented the value 6 twice and have obtained 7.

Common memory arbiters are found both on multiprocessor systems and on single processor systems whose I/O equipment is connected for *direct memory access* (DMA). Normally an I/O device would transfer each data word to the CPU for the CPU to store in the memory. However, this imposes an unacceptable overhead on the CPU. Instead, the I/O device is given the address of a block of memory. It only interrupts the CPU when the transfer of the whole block is completed. There is an arbiter to ensure that only one device (or the CPU) has access to the memory at any one time.

In this case we say that DMA is being implemented by *cycle stealing*. The memory is assumed to be driven at its maximum access speed, say one access per microsecond. Each such access is also called a memory cycle. To implement DMA the CPU is normally allowed to compute and access memory. When a data word arrives from an I/O device the right to access memory is usurped from the CPU and the device is allowed to "steal" a memory cycle. There is no real overhead. The memory cycle is needed anyway to store the word and with cycle stealing the CPU need not concern itself with individual words.

The computer hardware will be trusted to function properly. We only concern ourselves with the correctness of the system software. This is not always true of course and in practice one must be alert to hardware malfunction. One of the most spectacular bugs known to the Author was caused by a hardware fault that resulted in mixing two memory addresses instead of interleaving them. The net result was a store of data in the mixed-up address, and the presence of foreign data in these memory addresses was never explained by software specialists. Fortunately this sort of thing rarely happens.

Another way of using a common memory system is to define a primitive procedure call that is guaranteed to exclude other calls of the same procedure. That is, if two processes try to call the same procedure, only one will succeed and the losing process will have to wait. As usual it is not specified in which order simultaneous requests are granted.

In multiprogramming systems, the interrupt facility is used. A critical procedure is written as an interrupt routine to be executed when a process causes an interrupt. A hardware flag ensures mutual exclusion by inhibiting the interrupt—placing the computer in an uninterruptable state. Upon completion of the interrupt routine, the flag is reset and another process may now cause an interrupt.

Another method of implementing mutual exclusion is polling. Each process is interrogated in turn to see if it requires some service that must be done under mutual exclusion.

We shall allow ourselves the luxury of defining primitive instructions and, beyond the sketch in this section, we shall not worry about the implementation. With some experience in computer architecture and data structures it should not be too difficult to implement any of these primitives. However, the details differ widely from computer to computer. A study of the implementation kit may help. In the bibliography we give references to several descriptions of concurrent programming implementations. In addition, the serious student should study the architecture of whatever computer and operating system he is using.

2.6   CONCURRENT PROGRAMMING IN PASCAL-S

Sequential Pascal (and the subset used in this book) must be augmented by concurrent programming constructs. The concurrent processes are written as Pascal procedures and their identity as concurrent processes is established by their appearance in the **cobegin** . . . **coend** statement

$$\textbf{cobegin } P_1;\ P_2;\ \ldots;\ P_n \textbf{ coend.}$$

A request for concurrent execution of several processes may appear only in the main program and may not be nested. The semantics of the **cobegin** . . . **coend** statement are specified as follows:

> The **cobegin** statement is a signal to the system that the enclosed procedures are not to be executed but are to be marked for concurrent execution. When the **coend** statement is reached, the execution of the main program is suspended and the concurrent processes are executed. The interleaving of the executions of these processes is not predictable and may change from one run to another. When all concurrent processes have terminated, then the main program is resumed at the statement following the **coend**.

An additional notational device that we make use of is the statement **repeat** . . . **forever** which is exactly equivalent in its semantic content with **repeat** . . . **until** *false*. However, the latter is rather obscure and we prefer the more transparent notation.

The use of **repeat** ... **forever** emphasizes that these examples are intended to be prototypes of cyclic programs in operating systems and real-time systems.

To execute any of the examples in the book on the implementation kit you will generally have to do the following. (See Fig. 1.5 and, for a larger example, Fig. 4.18).

1. Replace **repeat** ... **forever** by a **for**-loop that will execute each process a fixed number of times or otherwise arrange for termination.

2. Insert *write* statements in the main program or in the concurrent processes to trace the execution of the program.

3. Often, certain procedures have been left unspecified to emphasize the generality of the algorithms. For example, in the producer–consumer problem, we have invoked procedures named *produce* and *consume*. These procedures must be specified. One possibility is to *produce* by incrementing an integer and *consume* by printing it.

4. *Warning* The interpreter in the kit is very inefficient, so do not get carried away with the size of the programs that you intend to execute.

2.7 SUMMARY

This list summarizes the concurrent programming abstraction.

1. A concurrent program will consist of two or more sequential programs whose execution sequences are interleaved.

2. The processes must be loosely connected. In particular, the failure of any process outside its critical section and protocols must not affect the other processes.

3. A concurrent program is correct if it does not suffer from violation of safety properties such as mutual exclusion and of liveness properties such as deadlock and lockout.

4. A concurrent program is incorrect if there exists an interleaved execution sequence which violates a correctness requirement. Hence it is sufficient to construct a scenario to show incorrectness; to show correctness requires a mathematical argument that the program is correct for all execution sequences.

5. No timing assumptions are made except that no process halts in its critical section and that, if there are ready processes, one is eventually scheduled for execution. We may impose other fairness requirements.

6. We shall extend our basic programming language with synchroniza-
    tion primitive instructions. As long as the syntax and semantics of
    these instructions are clearly defined we do not concern ourselves
    with their implementation.

2.8   EXERCISES

2.1   *Standing Exercise*   Write formal specifications of the programs in this book.
For example:
Specification for *sortprogram*.
Input: A sequence of 40 integers: $A=\{a_1, \ldots, a_{40}\}$.
Output: A sequence of 40 integers: $b=\{b_1, \ldots, b_{40}\}$.
Safety property: When the program terminates then (i) b is a permutation of a,
and (ii) b is ordered, i.e. for $1 <= i < 40$, $b_i <= b_i+1$.
Liveness property: The program terminates.

2.2   *Standing Exercise*   Test the example programs in the text.

3 THE MUTUAL EXCLUSION PROBLEM

3.1 INTRODUCTION

A solution to the mutual exclusion problem for two processes P_1 and P_2 will now be developed without introducing any primitive instructions (other than the common memory arbiter). The purpose of this chapter is not so much to present Dekker's elegant solution to this very difficult problem as to present Dijkstra's step-by-step development of Dekker's solution. During the development we will encounter most of the possible bugs that a concurrent program can have and thus illustrate the theoretical discussion of the previous chapter. The serious reader will want to try to analyze each attempted solution before reading further.

The mutual exclusion problem for two processes is as follows:
Two processes P_1 and P_2 are each executing in an infinite loop a program which consists of two sections, critical sections *crit*1 and *crit*2 and the

Fig. 3.1.

remainder of the program, non-critical sections *rem*1 and *rem*2. The execution of *crit*1 and *crit*2 must not overlap.

3.2   FIRST ATTEMPT

Let us imagine a "protocol igloo" containing a blackboard (Fig. 3.1). The igloo itself and the entrance tunnel are so small that only one person can be in the igloo at any given time. On the blackboard is written the number of the process whose "turn" it is to enter the critical section. The small size of the igloo is our metaphor for the memory arbiter.

A process wishing to enter its critical section crawls into the igloo when it is empty and checks the blackboard. If its number is written, it leaves the igloo and happily proceeds to its critical section. If the number of the other process is written then it sadly leaves the igloo to wait for the other process to finish.

Ideally, the unfortunate process would be able to take a nap until its turn arrives but we have no way of expressing this in an ordinary programming language. Instead, the process can run laps around the igloo to warm up (this is doing "nothing"). Periodically it can re-enter the igloo to check the blackboard. This is known as *busy waiting* because the energy expended by the waiting process is just purposeless work. When a process has completed its critical section, it writes the number of the other process on the board. Since only one process is in the igloo at any one time, the board shows either a one or a two (assuming that neither process malfunctions within the igloo).

```
program firstattempt;
var       turn: integer;
prodecure p₁;
begin
  repeat
    while turn=2 do (* nothing *);
    crit1;
    turn :=2;
    rem1
  forever
end;
procedure p₂;
begin
  repeat
    while turn=1 do (* nothing *);
    crit2;
    turn := 1;
    rem2
  forever
end;
```

begin (* *main program* *)
   turn := 1;
  cobegin
    p_1; p_2
  coend
end

Fig. 3.2.

This solution (Fig. 3.2) satisfies the mutual exclusion requirement. A process P_i enters its critical section only if *turn* = i. Since, by the common memory assumption, *turn* will be consistent (will have either a value of 1 or 2), only one process at a time enters its critical section. Since the value of *turn* is not changed until the termination of the critical section, a second process will not be able to infiltrate before termination of the first critical section.

Deadlock is also impossible. Since *turn* has either the value 1 or the value 2, exactly one process will always be able to progress or, to put it another way, it is impossible for both processes to be simultaneously stuck at the **while** loops.

Similarly, if every statement of each program takes a finite amount of time then the process whose turn it is to enter the critical section will always do so and will eventually allow the other process to enter also. Thus there is no lockout since neither process can prevent the other from entering its critical section.

Even though this solution fulfils our requirements for the correctness of concurrent programs, it does not fulfil one of the design requirements of the abstraction. The processes are not loosely connected. The right to enter the critical section is being explicitly passed from one process to the other.

This has two drawbacks. Firstly, if process P_1 needs to enter its critical section 100 times per day while P_2 needs to enter only once per day then P_1 is going to be coerced into working at P_2's pace of once per day. When P_1 has finished its critical section, *then*, it chalks up a 2 and until P_2 decides to re-enter the critical section, P_1 will be forced to wait.

The second drawback is as serious. Suppose P_1 is waiting for P_2 to execute a critical section and then change the number written on the board from 2 to 1. If by chance P_2 is devoured by a polar bear on its way to the igloo then not only is P_2 terminated but P_1 is hopelessly deadlocked. This is true even if P_2 is in *rem2* outside the critical section when it terminates, thus conforming with our assumption that a process may terminate only outside its critical section.

This explicit passing of control is a programming technique known as *coroutines*. By executing an instruction such as *resume* (p_2), P_1 is able to request that its execution be suspended in favor of P_2. P_2 then executes until it in turn returns the control of P_1 by a *resume* (p_1) statement. Coroutines are a useful programming technique but a system of coroutines must be designed as a single integrated process and are not a substitute for concurrent programs.

3.3 SECOND ATTEMPT

Fig. 3.3.

We try to remedy the previous solution by giving each process its own key to the critical section so if one is devoured by a polar bear then the other can still enter its critical section. There is now (Fig. 3.3) an igloo (global variable) identified with each process. It is worth noting that, while in the solution in Fig. 3.2 the variable *turn* is both read (Load) and written (Store) by both processes, the present solution may be easier to implement because each process reads but does not write the variable identified with the other process.

If P_1 (say) wishes to enter its critical section, it crawls into P_2's igloo periodically until it notes that c_2 is equal to 1 signifying that P_2 is currently not in its critical section. Having ascertained that fact, P_1 may enter its critical section after duly registering its entrance by chalking a 0 on its blackboard—c_1. When P_1 has finished, it changes the mark on c_1 to 1 to notify P_2 that the critical section is free.

```
program secondattempt;
var      c₁, c₂: integer;
procedure p₁;
begin
   repeat
```

```
      while c₂=0 do;
      c₁ := 0;
      crit1;
      c₁ := 1;
      rem1
    forever
  end;
  procedure p₂;
  begin
    repeat
      while c₁=0 do;
      c₂ := 0;
      crit2;
      c₂ := 1;
      rem2
    forever
  end;
  begin (* main program *)
    c₁ := 1;
    c₂ := 1;
    cobegin
      p₁; p₂
    coend
  end.
```

Fig. 3.4.

This program (Fig. 3.4) does not even satisfy the safety requirement of mutual exclusion. The following scenario gives a counter-example where the first column describes the interleaving and the next columns record the values of the variables.

	c_1	c_2
Initially	1	1
P_1 checks c_2	1	1
P_2 checks c_1	1	1
P_1 sets $c1$	0	1
P_2 sets $c2$	0	0
P_1 enters $crit1$	0	0
P_2 enters $crit2$	0	0

Since P_1 and P_2 are simultaneously in their critical sections, the program is incorrect.

3.4   THIRD ATTEMPT

```
program thirdattempt;
var      c₁, c₂: integer;
procedure p₁;
begin
  repeat
    c₁ := 0;
    while c₂=0 do;
    crit1;
    c₁ := 1;
    rem1
  forever
end;
procedure p₂;
begin
  repeat
    c₂ := 0;
    while c₁=0 do;
    crit2;
    c₂ := 1;
    rem2
  forever
end;
begin (* main program *)
  c₁ := 1;
  c₂ := 1;
  cobegin
    p₁; p₂
  coend
end.
```

Fig. 3.5.

Analyzing the failure of the second attempt, we note that, once P_1 has ascertained that P_2 is not in its critical section, P_1 is going to charge right into its critical section. Thus, the instant that P_1 has passed the **while** statement, P_1 is in effect in its critical section. This contradicts our intention that $c_1 = 0$ should indicate that P_1 is in its critical section because there may be an arbitrarily long wait between the **while** statement and the assignment statement.

The third attempt (Fig. 3.5) corrects this by advancing the assignment statement so that $c_1 = 0$ will indicate that P_1 is in its critical section even before it checks c_2. Hence P_1 is in its critical section the instant that the **while** has been successfully passed.

Unfortunately this program easily leads to system deadlock as seen by the following scenario:

	c_1	c_2
Initially	1	1
P_1 sets c_1	0	1
P_2 sets c_2	0	0
P_1 checks c_2	0	0
P_2 checks c_1	0	0

. . .

The continual checking of the variables can be continued indefinitely and cannot be considered progress. Thus the program is hopelessly deadlocked.

Even though this program is unacceptable because of the deadlock, it is instructive to prove that it satisfies the mutual exclusion property. By symmetry it is sufficient to show that: (P_1 in $crit1$) implies (P_2 is not in $crit2$).

1. (When P_1 entered $crit1$) then (c_2 was not 0).
   This follows from the structure of the program, namely the test on c_2 by P_1.
2. (c_2 is not 0) implies (P_2 is not in $crit2$).
   $crit2$ is bracketed between assignments to c_2 which ensure that this statement is always true.
3. (When P_1 entered $crit1$) then (P_2 was not in $crit2$).
   This is a logical consequence of (1) and (2).
4. (P_1 in $crit1$) implies (c_1 is 0).
   $crit1$ is bracketed between assignments to c_1.
5. (c_1 is 0) implies (P_2 does not enter $crit2$).
   The test will not allow P_2 through.
6. (P_1 in $crit1$) implies (P_2 does not enter $crit2$).
   A logical consequence of (4) and (5).
7. As long as (P_1 is in $crit1$), (P_2 will never enter $crit2$).
   This follows from (6). Since (6) refers to an arbitrary instant of time, then as long as its antecedent (P_1 in $crit1$) remains true, so will its consequent (P_2 does not enter $crit2$).
8. (P_1 in $crit1$) implies (P_2 is not in $crit2$).
   From (3) and (7).

Note that the proof has the simple structure of a deduction in the propositional calculus except for the need to express and deduce time-related properties such as "when", "as long as" etc. There is a formal logic called *temporal logic* that can express these properties and can be used to formally prove properties of concurrent programs. For example, the reasoning in this proof can be formalized as an induction on the time that has passed since P_1 entered $crit1$. We are trying to prove that mutual exclusion is never violated: (3) ensures the basis of the induction; (6) is an induction step: assuming that P_1 is now in $crit1$, we can deduce that P_2 will not now enter

$crit2$ so that upon the conclusion of the current instruction, mutual exclusion will still not be violated.

3.5 FOURTH ATTEMPT

```
program fourthattempt;
var c₁, c₂: integer;
procedure p₁;
begin
  repeat
    c₁ := 0;
    while c₂=0 do
      begin
      c₁ := 1;
      (* do nothing for a few moments *)
      c₁ := 0
      end;
    crit1;
    c₁ := 1;
    rem1
  forever
end;
procedure p₂;
begin
  repeat
    c₂ := 0;
    while c₁=0 do
      begin
      c₂ := 1;
      (* do nothing for a few moments *)
      c₂ := 0
      end;
    crit2;
    c₂ := 1;
    rem2
  forever
end;
begin (* main program *)
  c₁ := 1;
  c₂ := 1;
  cobegin
    p₁; p₂
  coend
end.
```

Fig. 3.6.

In the previous solution, when P_1 chalks up 0 on c_1 to indicate its intention to enter its critical section, it also turns out that it is insisting on its right to enter the critical section. It is true that setting c_1 before checking c_2 prevents the violation of mutual exclusion but if P_2 is not ready to yield then P_1 should yield.

In the next attempt (Fig. 3.6) we correct this stubborn behavior by having a process relinquish temporarily its intention to enter its critical section to give the other process a chance to do so. P_1 enters its igloo and chalks up a 0. If upon checking P_2's igloo, P_1 finds a 0 there too, it chivalrously returns to its igloo to erase the 0. After a few laps around the igloo it restores the signal $c_1 = 0$ and tries again. The comment is there simply to remind you that since arbitrary interleaving is permissible, the sequence of two assignments to the same variable is not meaningless.

First note that the previous proof of mutual exclusion holds here. From the above discussion, it should now be clear that there is such a thing as too much chivalry. If both processes continue yielding then neither will enter the critical section. The scenario is as follows:

	c_1	c_2
Initially	1	1
P_1 sets c_1	0	1
P_2 sets c_2	0	0
P_1 checks c_2	0	0
P_2 checks c_1	0	0
P_1 sets c_1	1	0
P_2 sets c_2	1	1
P_1 sets c_1	0	1
P_2 sets c_2	0	0

. . .

It is clear that this could be indefinitely extended and that liveness does not hold because neither process will ever enter its critical section. However, it is extremely unlikely ever to occur. Nevertheless we are forced to reject this solution. The main objection here is not so much that neither process will ever enter the critical section (it is unlikely that perfect synchronization continues indefinitely) but that we have no way of giving an *a priori* bound on the number of iterations that the loops will execute before they are passed. Thus we have no way of guaranteeing the performance of such a system.

Should this bug be classified as deadlock or lockout? On the one hand, both processes are looping on a protocol which is certainly not useful computation and the situation is similar to the previous attempt. However, we prefer to call this lockout to emphasize the following distinction. In the previous attempt the situation is hopeless. From the instant that the program is deadlocked, all future executions sequences remain deadlocked. In this case however, the slightest aberration of the scenario will free one of the processes and in practice this will eventually happen. The key notion here is the conspiracy between the processes and not the hopelessness of the situation. It is only because we wish to be able to guarantee a worst-case behavior that we reject the current attempt.

3.6 DEKKER'S ALGORITHM

```
program   Dekker;
var       turn: integer;
          c₁, c₂: integer;
procedure p₁;
begin
  repeat
    c₁ := 0;
    while c₂ = 0 do
      if turn = 2 then
        begin
          c₁ := 1;
            while turn=2 do;
          c₁ := 0
        end;
    crit1;
    turn := 2;
    c₁ := 1;
    rem1
  forever
end;
procedure p₂:
begin
  repeat
    c₂ :=. 0;
    while c₁ = 0 do
      if turn=1 then
        begin
          c₂ := 1;
          while turn=1 do;
          c₂ := 0
        end;
    crit2;
    turn := 1;
    c₂ := 1;
    rem2
  forever
end;
```

```
begin (* main program *)
    c₁ := 1;
    c₂ := 1;
    turn := 1;
    cobegin
        p₁; p₂
    coend
end.
```

Fig. 3.7.

Dekker's solution is an ingenious combination of the first and fourth attempted solutions. Recall that in the first solution we explicitly passed the right to enter the critical section between the processes. Unfortunately, the key to the critical section could be irretrievabley lost if one of the processes is terminated. In the fourth solution we found that keeping separate keys leads to the possibility of infinite deferment of one process to the other.

Dekker's algorithm (Fig. 3.7) is based on the previous solution but solves the problem of lockout by explicitly passing the right to *insist* on entering the critical solution. Each process has a separate igloo so it can go on processing even if one process is terminated by a polar bear. Note that we are here using the assumption that no process is terminated in its critical section (including the protocol).

There is now an "umpire" igloo with a blackboard labelled "*turn*", (Fig. 3.8). If P_1 chalks up a 0 on c_1 and then finds that P_2 has also chalked up a 0, it goes to consult the umpire. If the umpire has a 1 written upon it, then P_1 knows that it is its turn to insist and so P_1 periodically checks P_2's igloo. P_2 of course notes that it is its turn to defer and chivalrously chalks up a 1 on c_2 which will eventually be noted by P_1. P_2 meanwhile waits for P_1 to terminate

Fig. 3.8.

its critical section. Upon termination, P_1 not only frees the critical section by setting c_1 to 1 but also resets *turn* to 2 both to free P_2 from the inner loop and to transfer the right to insist to P_2.

Mutual exclusion is proved exactly as in Section 3.4 since the value of *turn* has no effect on the decision to enter the critical section.

Proving liveness is somewhat of a challenge. Symmetrically it is sufficient to prove that, if P_1 executes $c_1 := 0$ indicating its intention to enter the critical section, then eventually it does so. This is done in two parts. First we prove that if P_1 attempts to enter its critical section but cannot do so, eventually the variable *turn* is permanently held at the value 1. But if *turn* is held permanently at 1 then P_1 can always enter its critical section.

3.7   A PROOF OF DEKKER'S ALGORITHM

Let us now prove the liveness of Dekker's Algorithm. The algorithm is shown in flowchart form in Fig. 3.9. The formal statement that we want to prove is that if the program counter of process P_1 is at point α_2 (i.e. P_1 has left *rem*1 and thus expresses a wish to enter the critical section), eventually the program counter of P_1 will be at α_5 (i.e. P_1 may enter the critical section). Of course an exactly symmetrical proof will prove the liveness of P_2.

Our notation will be more concise: α_i will be an abbreviation for the statement that the program counter of P_1 is at α_i. Similarly for P_2 and β_i.

Remember the assumption that a process is never terminated in its critical section (including the protocols). Thus if P_1 is at α_6, it will eventually reach α_8. If P_1 is at α_3, it will eventually reach α_4 or α_5 though without further information we cannot specify which of the two statements will be reached. To simplify the proof, this assumption is extended to *remi* (α_1 and β_1). In the exercises we indicate the modifications needed if we allow P_1 to terminate in *remi*.

Note that if P_1 is at α_3 and $c_2 = 1$, we cannot conclude that eventually P_1 is at α_5. The assumption only guarantees that P_1 eventually tests the value of c_2; by then the value of c_2 could have been changed. To conclude that α_3 implies eventually α_5 we would have to show that α_3 and that the value of c_2 is held at 1 indefinitely. Thus, when by assumption the test is eventually done, the value of c_2 will in fact be 1.

Since the only assignments to c_i are in P_i, we can deduce the values of c_i from the α's and the β's, respectively. We express these facts as invariants, i.e. statements that are always true.

I1. $c_1 = 0$ if and only if α_3 or α_4 or α_5 or α_6 or α_7.

I2. $c_2 = 0$ if and only if β_3 or β_4 or β_5 or β_6 or β_7.

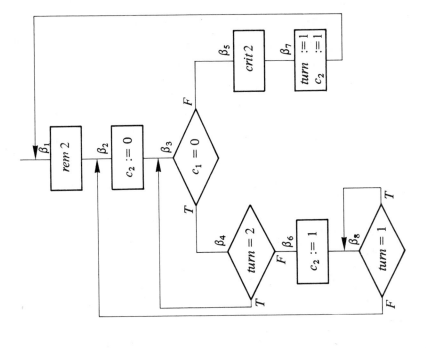

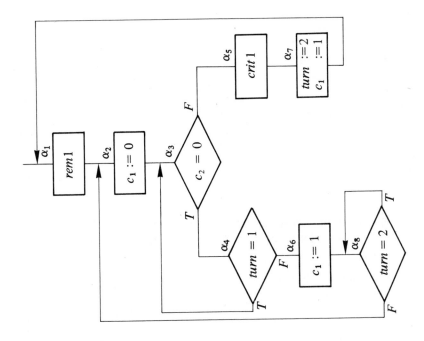

Fig. 3.9 Dekker's Algorithm. Initial values: $c_1 = c_2 = turn\ 1$.

Theorem   (α_2 and never α_5) is false, that is α_2 implies eventually α_5.

Proof

1. (α_2 and never α_5) and (*turn* held at 2) imply eventually (c_1 held at 1).
   Since (never α_5), P_1 eventually passes α_3 and thence to α_4. Since (*turn* held at 2), P_1 reaches α_6 and α_8 and is then blocked in the loop at α_8. By I1, as long as P_1 is at α_8, c_1 must equal 1.

2. (c_1 held at 1) and (*turn* held at 2) imply eventually (*turn* = 1).
   The truth of the two clauses concerning c_1 and *turn* together with the assumption that processes are not terminated implies that P_2 must eventually reach β_7 and assign the value 1 to *turn*.
   But (*turn* held at 2) and eventually (*turn* = 1) means that there will be a point of time when *turn* is simultaneously 1 and 2. This contradicts the consistency of the values in the common memory. From (α_2 and never α_5) and (*turn* held at 2) we have deduced a contradiction. Thus it must be that (α_2 and never α_5) implies eventually (*turn* is not 2). Since *turn* = 2 or *turn* = 1 we have proved.

3. (α_2 and never α_5) implies eventually (*turn* = 1).

4. (α_2 and never α_5) implies (never α_5) and (never α_7) and (never α_1).
   The only way to reach α_1 or α_7 from α_2 is to pass through α_5. But we assume that we never reach α_5.

5. (α_2 and never α_5) implies eventually (*turn* held at 1).
   Once the value of *turn* is 1 (as ensured by (3)) the only way that the value can change back to 2 is to execute α_7. By 4 this will never happen.

6. (α_2 and never α_5) implies eventually (P_1 loops forever at α_3–α_4).
   By (4) and (5), eventually [(*turn* held at 1) and (P_1 is never at α_5)]. Hence since P_1 must reach α_3–α_4 from α_2, α_6 or α_8 it will then loop forever at α_3–α_4.

7. (α_2 and never α_5) implies eventually (c_1 held at 0).
   By (6), eventually we loop at α_3–α_4 which implies by I1 that c_1 will be held at 0.

8. (c_1 held at 0) and (*turn* held at 1) imply eventually (c_2 held at 1).
   Similar to (1): P_2 must eventually loop at β_8. Then by I2, c_2 is held at 1.

9. (α_2 and never α_5) implies eventually (c_2 held at 1).
   From (5), (7) and (8).
   But (8) contradicts (6): if c_2 is held at 1 then P_1 cannot be looping

at $\alpha_3-\alpha_4$. From (α_2 and never α_5) we have deduced a contradiction. Thus (α_2 and never α_5) is false.

3.8 CONCLUSION

Mutual exclusion of two processes is about the simplest problem in concurrent programming. The difficulty of obtaining a correct solution to such a simple problem suggests that programming features more powerful than the common memory arbiter will be needed. In the exercises you can explore some other solutions of the type given here.

In particular, the solutions of the mutual exclusion problem for n processes are so difficult that they are of more or less academic interest only, especially when compared with the trivial solution to the problem given, say, by semaphores.

There is another defect in the common memory arbiter and that is the busy wait that is used to achieve synchronization. The solutions all contain a statement: **while** *condition* **do** (* *nothing* *). Unless you have a dedicated computer doing the looping this is a waste of CPU computing power. Even if there is no CPU waste (as would be the case if the processes were I/O controllers) there is the severe overhead associated with cycle stealing. Thus the frequent accesses to *turn* in Dekker's solution can prevent useful computation from being done by other processes.

The primitives discussed in the next chapters uniformly suspend the execution of the blocked processes. This is usually implemented by keeping a queue of processes, i.e. a queue of small blocks of memory containing essential information on the blocked processes. Thus the overhead is only a small amount of memory and the small amount of computation needed to manage the queue.

A final objection to Dekker's algorithm is that it uses a common variable which is written into by both processes. In the exercises we discuss Lamport's algorithms which have the advantage that each variable need only be written by one process. Thus his algorithms are suitable for implementation on distributed systems where the values of the variables can be transmitted and received but where each variable is written into only on the computer in which it physically resides.

3.9 EXERCISES

3.1 (Dijkstra) Fig. 3.10 is a solution to the mutual exclusion problem for n processes that is a generalization of Dekker's solution.

(a) *Show that mutual exclusion holds.
(b) Show that deadlock does not occur.
(c) Show that lockout is possible.

```
program   Dijkstra;
const     n = . . . ; (* number of processes *)
var       b, c: array [0 . . n] of boolean;
          turn: integer;
procedure process(i : integer);
var       j: integer;
          ok: boolean;
begin
  repeat
    b[i] := false;
    repeat
      while turn <> i do
        begin
            c[i] := true;
            if b[turn] then turn := i
        end;
      c [i] := false;
      ok := true;
      for j := 1 to n do
        if j <> i then
        ok := ok and c[j]
    until ok;
    crit;
    c[i] := true; b[i] := true;
    turn := 0;
    rem
  forever
end;
begin (* main program *)
  for turn := 0 to n do
    begin
      b[turn] := true;
      c[turn] := true
    end;
  turn := 0;
  cobegin
    process(1);
    process(2);
    . . .
    process(n)
  coend
end.
```

Fig. 3.10.

3.2 (Lamport) Fig. 3.11 is (what the Author calls) the Dutch Beer version of the Bakery Algorithm, restricted to two processes.

(a) Show safety and liveness. (*Hint* The variables are supposed to represent "ticket" numbers". The process with the lower ticket number enters its critical section. In case of a tie, it is arbitrarily resolved in favor of P_1.)

(b) Show that the commands $n_i := 1$ are necessary.

(c) *Extend the algorithm to n processes. (*Hint* Each process will choose a ticket number greater than the maximum of all outstanding ticket numbers. It will then wait until all processes with lower numbered tickets have completed their critical sections.)

```
program   dutchbeer;
var       n₁, n₂: integer;
procedure p₁;
begin
  repeat
    n₁ := 1;
    n₁ := n₂+1;
    while (n₂ <> 0) and (n₂ < n₁) do;
    crit1;
    n₁ := 0;
    rem1
  forever
end;
procedure p₂;
begin
  repeat
    n₂ := 1;
    n₂ := n₁+1;
    while (n₁ <> 0) and (n₁ <= n₂) do;
    crit2;
    n₂ := 0;
    rem2
  forever
end;
begin (* main program *)
  n₁ := 0;
  n₂ := 0;
  cobegin
    p₁; p₂
  coend
end.
```

Fig. 3.11.

3.3 Fig. 3.12 is Lamport's Bakery Algorithm restricted to two processes.

   (a) Show the safety and liveness of this solution.

   (b) Generalize to n processes.

   (c) Show that for $n > 2$ the values of the variables n_i are not bounded.

   (d) *Suppose we allow a read (i.e. Load) of a variable n_i to return any value if it takes place simultaneously with a write (Store) of n_i by the ith process. Show that the correctness of the algorithm is not affected. Note, however, that we require the write to execute correctly. Similarly, all reads which do not overlap writes to the same variable must return the correct values.

   (e) Show that the correctness of the Dutch Beer version of the algorithm is not preserved under the malfunction described in (d).

```
program  bakery;
var       c₁, c₂, n₁, n₂: integer;
procedure p₁;
begin
  repeat
    c₁ := 1;
    n₁ := n₂+1;
    c₁ := 0;
    while c₂ <> 0 do;
    while (n₂ <> 0) and (n₂ < n₁) do;
    crit1;
    n₁ := 0;
    rem1
  forever
end;
procedure p₂;
begin
  repeat
    c₂:= 1;
    n₂ := n₁+1;
    c₂ := 0;
    while c₁ <> 0 do;
    while (n₁ <> 0) and (n₁ <= n₂) do;
    crit2;
    n₂ := 0;
    rem2
  forever
end;
begin (* main program *)
  c₁ := 0;
  c₂ := 0;
  n₁ := 0;
  n₂ := 0;
  cobegin
    p₁; p₂
  coend
end.
```

Fig. 3.12.

3.4 Fig. 3.13 is a solution to the mutual exclusion problem for two processes. Discuss the correctness of the solution: if it is correct, then prove it. If not, write scenarios that show that the solution is incorrect.

Several sychronization primitives that have been used are based on hardware instructions that enable several assignment statements to be executed as one (indivisible) primitive instruction. The solutions to the mutual exclusion problem using these primitives are very simple, but they remain busy wait algorithms in contrast to algorithms using the primitives to be studied in the next chapter.

```
program  attempt;
var       c₁, c₂: integer;
procedure p₁;
begin
  repeat
    rem1;
    repeat
      c₁ := 1 - c₂
    until c₂ <> 0;
      crit1;
      c₁ := 1
  forever
end;
procedure p₂;
begin
  repeat
    rem2;
    repeat
      c₂ := 1 - c₁
    until c₁ <> 0;
      crit2;
      c₂ := 1
  forever
end;
begin (* main program *)
  c₁ := 1;
  c₂ := 1;
  cobegin
    p₁; p₂
  coend
end.
```

Fig. 3.13.

3.5 The IBM 360/370 computers have an instruction called TST (Test and Set). There is a system global variable called c (Condition Code). Executing $TST(l)$ for local variable l is equivalent to the following two assignments:

$$l := c;$$
$$c := 1.$$

(a) Discuss the correctness (safety, deadlock, lockout) of the solution of the mutual exclusion problem shown in Fig. 3.14.
(b) Generalize to n processes.
(c) What would happen if the primitive TST instruction were replaced by the two assignments?
(d) *Modify the implementation kit to include the TST instruction.

```
program   testandset;
var       c: integer;
procedure p₁;
var       l: integer;
begin
  repeat
    rem1;
    repeat
      TST(l)
    until l = 0;
      crit1;
      c := 0
  forever
end;
procedure p₂;
var       l: integer;
begin
  repeat
    rem2;
    repeat
      TST(l)
    until l = 0;
      crit2;
      c := 0
  forever
end;
begin (* main program *)
  c := 0;
  cobegin
    p₁; p₂
  coend
end.
```

Fig. 3.14.

3.6 The EX instruction exchanges the contents of two memory locations. $EX(a,b)$ is equivalent to an indivisible execution of the following assignment statements:

$$temp := a;$$
$$a := b;$$
$$b := temp.$$

(a) Discuss the correctness (safety, deadlock, lockout) of the solution for mutual exclusion shown in Fig. 3.15.

(b) Generalize to n processes.

(c) What would happen if the primitive EX instruction were replaced by the three assignments?

(d) *Modify the implementation kit to include the EX instruction.

```
program   exchange;
var        c: integer;
procedure p₁;
var        l: integer;
begin
  l := 0;
  repeat
    rem1;
    repeat
      EX (c,l)
    until l = 1;
      crit1;
      EX (c,l)
  forever
end;
procedure p₂;
var        l: integer;
begin
  l := 0;
  repeat
    rem2;
    repeat
      EX (c,l)
    until l = 1;
      crit2;
      EX (c,l)
  forever
end;
begin (* main program *)
  c := 1;
  cobegin
  p₁; p₂
  coend
end.
```

Fig. 3.15.

3.7 *If we allow P_2 to terminate in *rem2*, what changes need to be made in the proof of the liveness of Dekker's algorithm? (*Hint* If P_2 is terminated in *rem2*, then it is true by I2 that c_2 is held at 1).

4 SEMAPHORES

4.1 INTRODUCTION

The scientific study of concurrent programs was given a decisive thrust with the introduction of the semaphore by Dijkstra. Semaphores are easy to implement and yet sufficiently powerful that they can be used to give elegant solutions to concurrent programming problems. They can be used to define or implement more powerful structured primitives.

A *semaphore s* is an integer variable which can take on only non-zero values. Once s has been given its initial value, the only permissible operations on s are to call the procedures *wait(s)* and *signal(s)* which are primitive operations (the original notation is $P(s)$ for *wait(s)* and $V(s)$ for *signal(s)*, which are the first letters of the corresponding words in Dutch). The definition of these operations is as follows:

wait(s): If $s > 0$ then $s := s - 1$ else the execution of the process that called *wait(s)* is suspended.

signal(s): If some process P has been suspended by a previous *wait(s)* on this semaphore s then wake up P else $s := s + 1$.

Remark 1   If the semaphore only assumes the values 0 and 1, it is called a *binary semaphore*. A semaphore which can take arbitrary non-negative integer values is called a *general semaphore*.

Remark 2   *wait* and *signal* are the only operations allowed. In particular, assignments to s or tests of the value of s are prohibited except for an assignment to s of an initial non-negative value in the main program. (The implementation kit does not enforce this restriction nor does it distinguish binary from general semaphores.)

Remark 3   We have defined the procedures as primitive operations. This means that they exclude one another just as Load and Store to the same memory word exclude one another. (Semaphore operations on distinct semaphores need not exclude one another). Hence if a *wait* and a *signal* (or two *wait*'s or two *signal*'s) occur simultaneously they are executed one at a time though we do not know in what order they are executed.

Remark 4 The definition of *signal* does not specify which process is woken if more than one process has been suspended on the same semaphore. For the purpose of constructing a scenario you may assume that any process you wish is in fact selected. A FIFO semaphore which maintains a first-in, first-out queue of suspended processes could also be defined. There are more sophisticated definitions of fairness but that is a slightly more advanced topic and we suggest that you skip it on the first reading.

(*Advanced*) *Remark* 5 The classical (busy-wait) definition of semaphores is even weaker:

wait(s): When $s > 0$ then $s := s - 1$.

Read this as: whenever the value of s is greater than zero, the process may execute $s := s - 1$ and continue.

signal(s): $s := s + 1$.

Suppose that process P executes *signal(s)* when the queue contains the blocked processes $Q_1, \ldots, Q_k$. This definition does not guarantee that any of the $Q_1, \ldots, Q_k$ are woken. The interleaving of execution sequences allows any process (even P) to execute *wait(s)* and decrement s before any of $Q_1, \ldots, Q_k$ have a chance to do so.

Another definition is that of the strongly fair semaphore which is assumed to have the following property:

If P is suspended on s and s becomes greater than zero infinitely often then eventually *signal* will choose to wake P.

(End of Remark 5).

4.2 MUTUAL EXCLUSION

Figure 4.1 is a solution to the mutual exclusion problem using semaphores.

```
program   mutualexclusion;
var       s: (* binary *) semaphore;
procedure p₁;
begin
  repeat
    wait (s);
    crit1;
    signal(s);
    rem1
  forever
end;
```

```
procedure p₂;
begin
  repeat
    wait(s);
    crit2;
    signal(s);
    rem2
  forever
end;
begin (* main program *)
  s := 1;
  cobegin
    p₁; p₂
  coend
end.
```

Fig. 4.1.

All the synchronization is veiled by the powerful features of the semaphore. That is the way it should be. Language features are defined; compiler writers and systems programmers figure out how to implement it; and everyone else uses the feature with a reasonable assurance that it will work.

Let us examine this algorithm in detail with the aid of the igloo model. Our igloo (Fig. 4.2) now has in addition to its blackboard a deep-freezer. A process enters and performs a *wait*: if there is a 1 on the board it can enter its critical section; otherwise, it goes into hibernation in the freezer. Note that once a process enters the freezer it has cleared the interior of the igloo and

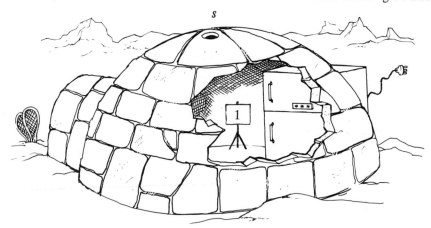

Fig. 4.2.

another process can enter even though the first process has not actually "completed" the execution of the *wait*. As an implementation detail, the freezer (= queue of suspended processes) must be large enough to contain the number of processes in the system or at least the number of processes that may be waiting on the semaphore *s*.

Upon completion of the critical section, the signalling process enters the igloo and releases a process from the freezer. If there are no such processes, it simply chalks up a one to indicate that the critical section is free. In the case of a binary semaphore, a signalling process will always find a zero on the board (why?). If the blackboard of a general semaphore has any non-zero number written on it, the process can deduce that the freezer is empty (why?).

This solution to the mutual exclusion problem is very similar to one of our earliest attempts in the previous chapter in which we had the processes pass the key to the critical section back and forth. We are saved from trouble here by the fact that the testing of *s* and the setting of *s* to zero are encapsulated in one primitive instruction. Thus if P_1 notes that *s* is 1, it will set *s* to 0 before P_2 has a chance to test the value of *s*.

Mutual exclusion and absence of deadlock are easy to show from the following property of the program: *s* will have the value zero if and only if exactly one process is in its critical section. This can be formally proved as follows. Consider the value of the expression $E=s+the$ *number of processes in a critical section*. Certainly $E=1$ at the start of the concurrent program since $s=1$ and no process is in a critical section.

Now use the following inductive argument. Assume that $E=1$ at any point in any interleaved execution sequence. The execution sequence continues by choosing to execute either a step of P_1 or a step of P_2. We argue by inspecting the program that in any case the truth of $E=1$ is preserved. Hence by induction, $E=1$ is always true because any execution sequence is constructed starting from the initial state by successively choosing either a step of P_1 or of P_2. For example, if $E=1$ because there are no processes in the critical section and $s=1$, P_1 can choose to enter the critical section by executing *wait(s)*. It leaves $s=0$ and *number of processes in the critical section* $=1$, i.e. $E=1$.

The formula $E=1$ is called an invariant of the computation. Invariants are proved by induction. The initial state of the computation satisfies the invariant and every transition between possible states of the computation preserves the truth of the invariant. To prove this, we assume the truth of the invariant as an induction hypothesis and then check that the invariant is still true in the new state resulting from the transition.

Note that, even in this simple case, the full proof is rather tedious since the induction step must be proved for every pair: (location of P_1's program counter, location of P_2's program counter). Of course, most steps are trivial.

The only ones needing any reasoning are executions of the semaphore instructions.

We are now in a position to prove the liveness of the solution.

Theorem (P_1 wishes to enter $crit1$) implies eventually (P_1 enters $crit1$).

Proof

1. (P_1 wishes to enter $crit1$) implies eventually (P_1 enters $crit1$) or (P_1 is indefinitely suspended because $s=0$).
2. (P_1 is indefinitely suspended because $s=0$) implies (P_2 is in $crit2$).
   This follows from the invariant $E=1$ and the fact that there are only two processes P_1 and P_2.
3. (P_2 is in $crit2$) implies eventually (P_2 executes $signal(s)$).
   We assume that no process is terminated in its critical section.
4. (P_2 executes $signal(s)$) and (P_1 is indefinitely suspended because $s=0$) implies (P_1 enters $crit1$).
   See the definition of the $signal$ operation.
5. (P_1 wishes to enter $crit1$) implies eventually (P_1 enters $crit1$).
   The possibility that P_1 is indefinitely suspended on $s=0$ has led to a contradiction.

Remark Note that in (4) we have tacitly used the fact that there are only two processes. Otherwise we could not prove that P_1 enters its critical section and not some other process.

(Advanced) Remark (4) of course does not hold under the busy-wait definition of semaphores. In fact lockout is possible under that definition if the signalling process executes another *wait* before the suspended process notes that $s > 0$. However, fair semaphores are sufficient to prove absence of lockout though a proof such as ours would have to be invoked inductively so that eventually some *signal* would in fact wake P_1.
(End of Advanced Remark).

The mutal exclusion problem for n processes is solved by the identical program (Fig. 4.3).

```
procedure mutualexclusion;
const n = . . . ; (* number of processes *)
var    s: (* binary *) semaphore;
procedure process(i: integer);
begin
  repeat
    wait(s);
    crit;
```

```
    signal(s);
    rem
  forever
end;
begin (* main program *)
  s := 1;
  cobegin
    process(1);
    process(2);
      . . .
    process(n)
  coend
end.
```

Fig. 4.3.

In this case lockout is a possibility. Suppose there are three processes and that the arbitrary choice of a process woken by *signal* is such that the process with the lowest index is always chosen. Then P_3 could be indefinitely delayed as P_1 and P_2 conspire to wake each other up.

(*Advanced*) *Remark* Morris has found a lockout-free solution to the mutual exclusion problem for *n* processes. The solution is very complex and uses additional variables and semaphores to set a limit on the size of a "batch" of processes that may simultaneously wait on the semaphore used for mutual exclusion. Since every process eventually completes its critical section and leaves the batch, all these processes will be eventually processed before a new batch is allowed to compete for mutual exclusion. The solution depends on the definition of semaphores that requires a signalling process to wake a waiting process.
(End of Advanced Remark).

Consider the problem of allowing at most *k* out of the *n* processes to simultaneously access the critical section. For example, a computer with two printers could allow two jobs to be printed simultaneously. The only change needed is to initialize the semaphore to *k*. *k* processes will successfully decrement *s* until its value is zero. Then a process must wait until one of these *k* processes completes its critical section and signals. We can see here that it is necessary for *signal* also to be a primitive instruction. Otherwise if $k=3$, $s=1$ and two processes leave their critical sections simultaneously then the concurrent execution could leave $s=2$ even though all processes are now out of their critical sections.

Can this be done using binary semaphores only? The answer is yes, but the solution is difficult. In the Exercises you can study the solution found by Kessels and Martin (1979).

4.3 THE PRODUCER–CONSUMER PROBLEM

Along with mutual exclusion, the producer–consumer problem is an abstraction of applications of concurrent programming that are found throughout operating systems. We choose to introduce this problem here because the semaphore has a natural interpretation in terms of the producer–consumer problem and because Dijkstra's presentation offers another elegant series of solutions to a concurrent programming problem.

The producer–consumer problem arises because the producer of data must have somewhere to store it until the consumer is ready and the consumer must not try to consume data that is not there. It is perfectly valid to require that a rendezvous between the two must take place. Then the producer produces if and only if the consumer is ready to consume. If either process arrives early then it is required to wait. The rendezvous is the reasonable thing to do unless the two processes have a common memory. It is the basis of the Ada synchronization primitives.

If, however, the data rates of the producer or the consumer vary during the execution of the program then buffering is necessary. An example is the type-ahead feature found on most terminal systems. The user is allowed to produce several commands without waiting for the computer to consume each command. Similarly, the computer may produce more information than can be conveniently presented on the terminal screen. A buffer is used to average out such peak data rates.

A *buffer* is a segment of memory common to both the producer and the consumer. If the buffer is large enough to handle peaks of data production, both producer and consumer maintain a steady high average rate of data transfer without fearing a malfunction because of occasional peaks. The operation of a buffer is the same as that of a shock absorber in a car which stores a peak of energy and then releases it slowly so that both the car and its occupants can tolerate this input of energy. If a bump is hit which is beyond the capacity of the shock absorber or if the bumps are produced at a rate too rapid to allow the shocks to be consumed by a slow release of energy, then the result is unfortunate for the owner of the car.

A related use of buffers is to accommodate I/O equipment that accepts only aggregates of data. Disks and tapes can read and write only blocks of

```
repeat
    produce record v;
    b[in] := v;
    in := in+1
forever;
```

Fig. 4.4.

data. Many terminal systems require that complete messages be transmitted instead of individual characters. Even though both the producer and the consumer may be working at the same average rate, the artificial imbalance caused by blocking the data requires that buffering be used.

For now, let us assume that we have an infinite buffer. In programming notation this can be expressed as an infinite array: $b[0], b[1], \ldots$. The producer can then simply pour his data into the buffer (Fig. 4.4) (in is a global variable that counts the number of records produced). The consumer on the other hand must assure that it is not consuming from an empty buffer (Fig. 4.5) (out is a variable that counts the number of records consumed). Initially we set $in = out = 0$. Note that we have abstracted away many details of the actual buffering process, in particular the structure of the records and the processing to be done with them. However, the main idea of buffering is captured.

repeat
    wait until in > out;
    $w := b[out];$
    $out := out + 1;$
    consume record w
forever;

Fig. 4.5.

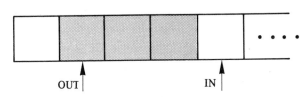

OUT   IN

Fig. 4.6.

Let $s = in - out$ (Fig. 4.6). s is then the number of records in the buffer. What values can s take? s is initially 0. s can increase and then decrease arbitrarily except that if s reaches 0 then the consumer will refuse to reduce s below 0. Instead it waits until the producer places another value in the buffer. If we arrange for the producer to force the consumer into immediate consumption of this new value then $0 = in - out = (in + 1) - (out + 1)$.

s behaves like a semaphore. In fact the statement *wait until in > out* in the consumer can easily be implemented by *wait(s)* assuming that *signal(s)* is added to the producer to wake up the consumer. Thus a solution to the producer–consumer problem can be written as shown in Fig. 4.7 (where we have further abstracted the buffer manipulations by *append* (to buffer) and *take* (from buffer)).

```
program  producerconsumer;
var      n: semaphore;
procedure producer;
begin
  repeat
    produce;
    append;
    signal(n)
  forever
end;
procedure consumer;
begin
  repeat
    wait(n);
    take;
    consume
  forever
end;
begin (* main program *)
  n := 0;
  cobegin
    producer; consumer
  coend
end.
```

Fig. 4.7.

Thus the semaphore can be viewed as counting the difference between the number of signals sent by *signal* and the number of signals received by *wait*. It is the mere fact of signalling that is being counted and not the content of the signal. A semaphore can be implemented by a message passing system though it is wasteful to use the fixed size message element to transmit a null message.

4.4   MORE ON THE PRODUCER–CONSUMER PROBLEM

We now describe a series of solutions to the producer–consumer problem under a different hypothesis than in the previous section. Let us assume that the statements *append* and *take* are critical sections that must not overlap.

(*Advanced*) *Remark*   A buffering system may be implemented as a chain of small buffers linked together. Obtaining or releasing a small buffer is usually a critical section to ensure the consistency of the pool of free buffers. Another common situation that requires mutual exclusion is the case of multiple producers or consumers. For example, all the terminals in a transaction processing system might place their request in a single queue;

this would be an instance of multiple producers. If a second processor were added to the system to improve its performance, we would have multiple consumers.
(End of Advanced Remark).

In the program in Fig. 4.8, a (binary) semaphore s is used in addition to the general semaphore n to achieve mutual exclusion.

```
program   producerconsumer;
var       n: semaphore;
          s: (* binary *) semaphore;
procedure producer;
begin
  repeat
    produce;
    wait(s);
    append;
    signal(s);
    signal(n)
  forever
end;
procedure consumer;
begin
  repeat
    wait(n);
    wait(s);
    take;
    signal(s);
    consume
  forever
end;
begin (* main program *)
  n := 0;
  s := 1;
  cobegin
    producer; consumer
  coend
end.
```

Fig. 4.8.

Suppose that a programming bug was made and that instead of *signal(s); signal(n)* was written *signal(n); signal(s)*. This shouldn't affect the safety of the solution because the solution must be safe even if the interleaving is such that the two *signal*'s are executed successively with no intervening statements from other processes. It is conceivable that the liveness could be affected since the release of one waiting process out of turn could allow it to conspire against other processes. There is an example of this in the lockout-free algorithm of Morris mentioned earlier. In this case, fortunately, such a bug does not affect the liveness. The consumer must wait on both semaphores before consuming. Since a *signal* is never blocked and there is only one waiting process, it does not matter in which order the *signal*s are issued. If the consumer is released from the *wait(n)* by *signal(n)* it will still be prevented from *take*ing prematurely by *wait(s)*.

On the other hand, exchanging the *wait*s is fatal. Consider the following simple scenario: the consumer executes *wait(s)* which is successful (because *s* is initially 1) and then it is blocked by *wait(n)* (because *n* is initially 0). But now *s* is 0 and the producer will never be able to *append* to the buffer. Thus the system is deadlocked.

This shows up a serious weakness of semaphores. There is no way to conditionally enter or leave a *wait*; nor is there a way to examine the value of the semaphore without executing a *wait* and becoming vulnerable to being blocked. A more powerful primitive which does not have this weakness is the *conditional critical region*. This region evaluates an expression on the ordinary program variables unlike the semaphore variable which is not freely accessible. For the time being, we continue our discussion of the low level semaphore and later we return to more powerful primitives.

An advantage of semaphores is that they are easy to implement; in particular, binary semaphores are simple to implement because we do not have to worry about the maximum value that needs to be provided for as is the case with the general semaphore. One bit is enough. The next solution (Fig. 4.9) to the producer–consumer problem is by binary semaphores only. Of course we will need an integer variable *n* to count the number of elements in the buffer since that information will no longer be stored in the semaphore. The semaphore *delay* will block the consumer if the buffer is empty.

```
program    producerconsumer;
var        n: integer;
           s: (* binary *) semaphore;
           delay: (* binary *) semaphore;
procedure producer;
begin
  repeat
    produce;
    wait(s);
    append;
    n := n+1;
    if n=1 then signal(delay);
    signal(s)
  forever
end;
```

```
procedure consumer;
var    m: integer; (* a local variable *)
begin
   wait(delay);
   repeat
      wait(s);
      take;
      n := n-1;
      m := n;
      signal(s);
      consume;
      if m=0 then wait(delay)
   forever
end;
begin (* main program *)
   n := 0;
   s := 1;
   delay := 0;
   cobegin
      producer; consumer
   coend
end.
```

Fig. 4.9.

Note the initial *wait(delay)* so that the consumer does not begin to execute while the buffer is empty. Also, if the processes are running at more or less the same speed, neither is ever blocked on the semaphore *delay*. This is because *wait(delay)* is executed only if the buffer is emptied which need not occur frequently. The *wait(delay)* has been taken out of the *wait(s)* . . . *signal(s)* bracket to avoid the previously discussed deadlock.

The new feature in the solution is the use of the local variable m to allow the consumer to test the value of n as it was inside the criticial section. If the statement in the consumer had read: **if** $n=0$ **then** *wait(delay)* then the following scenario shows that a superfluous signal can occur which leads to consumption from an empty buffer—a flagrant breach of safety. Define a cycle of the producer (consumer) as execution of the statements of the producer (consumer) process from one occurrence of the *produce (consume)* to the next.

In the line marked (*) the consumer has skipped the *wait* in the statement **if** $n=0$ **then** *wait(delay)* because even though it noted that $n=0$, the producer has meanwhile incremented n. The notation $n=-1$ means that the consumer has just consumed an element that is not there: $-1=n=in-out$ so $out=in+1$, i.e. the consumer has consumed its $(in+1)$st element while the producer has produced only in elements.

Action	*n*	*delay*
Initially	0	0
Producer cycles	1	1
Consumer executes to *consume*	0	0
Producer cycles	1	1
Consumer cycles	0	1(*)
Consumer cycles	−1	0

Using the local variable *m*, this bug will not occur (Check!). It is true that the consumer is making a decision based on stale information: it could be the case that, meanwhile, the producer has produced a new element. Then the consumer will execute *wait(delay)* and immediately pass it if the producer has already signalled. The overhead of a superfluous *wait* is certainly preferable to a violation of safety.

4.5   THE SLEEPING BARBER

There is another slight improvement that can be made to this program. The point of this discussion is not so much the improvement itself which may or may not be significant. What is interesting is how a careful analysis of the synchronization requirement in a problem can lead to a different and better solution. The moral of the story will be that before you decide to wait on a semaphore, you must clearly understand what you are waiting for.

Suppose we have the (common) case where the producer and consumer are running at roughly the same speed. The scenario could be:

Producer: *append*; *signal*; *produce*; . . . ; *append*; *signal*; *produce*; . . .
Consumer: *consume*; . . . ; *take*; *wait*; *consume*; . . . ; *take*; *wait*; . . .

The producer always manages to append a new element to the buffer and signal during the consumption of the previous element by the consumer. This is not unreasonable since the processing to be done with the data is assumed to be significant compared with the buffer manipulation and the synchronization. The producer is always appending to an empty buffer and the consumer is always taking the last element in the buffer; hence the execution of *signal* and *wait* on every cycle. Thus even though the consumer will never block on the semaphore, the processes nevertheless are executing a large number of calls to the semaphore mechanism which does involve non-negligible overhead.

In the program in Fig. 4.10, we allow *n* to have the value −1 which is to mean that not only is the buffer empty but that the consumer has detected this fact and is going to block until the producer supplies fresh data.

```
program   sleepingbarber;
var       n: integer;
          s: (* binary *) semaphore;
          delay: (* binary *) semaphore;
procedure producer;
begin
  repeat
    produce;
    wait(s);
    append;
    n := n+1;
    if n=0 then signal(delay);
    signal(s)
  forever
end;
procedure consumer;
begin
  repeat
    wait(s);
    n := n-1;
    if n=-1 then
      begin
        signal(s);
        wait(delay);
        wait(s)
      end;
    take;
    signal(s);
    consume
  forever
end;
begin (* main program *)
  n := 0;
  s := 1;
  delay := 0;
  cobegin
    producer; consumer
  coend
end.
```

Fig. 4.10.

Before the consumer waits on *delay* it is careful to release mutual
exclusion by signalling *s* to allow the producer to append a new element to

the buffer. The structure of the consumer shows an alternative programming construct to the use of a local variable as in the previous program. The test on *n* is made inside the critical section and the waiting outside. Since the test is made inside, there is no chance for the producer to change the value of *n* between the decrement of *n* and the test.

Once the *delay* is completed we are careful to ask for the return of mutual exclusion, hence the extra *wait(s)*. If we try the scenario sketched at the beginning of this section we find that even if there was only a single element in the buffer, after $n := n - 1$ the value of *n* will be zero so the consumer falls through to *consume*. If the producer can append a new element fast enough, the consumer need never execute the *wait(delay)*.

Fig. 4.11.

The two solutions can be illustrated by the model of the Sleeping Barber. A barber has a two-room shop as shown in Fig. 4.11. One room with the barber-chair and a waiting room. The three doors shown are: from the street to the waiting room, from the waiting room to the chair and from there back to the street. The doors are assumed to be narrow and allow at most one person to pass at a time (the common memory arbiter). Let us now model the stream of customers as a data elements which the barber must "consume". The waiting room is the buffer. We now write algorithms for both the barber and the customers.

Algorithm 4.1

Barber When you have finished with a customer, show him out and check the waiting room. If there is a customer, escort him to the chair; otherwise, go to sleep in the chair.

Customer When you enter the waiting room: if there are other customers then join them. If not, open the door to see if the barber is busy; if so, close the door and wait your turn. If the barber is asleep then wake him.

In Algorithm 4.1, if the rate at which customers enter matches the rate of the barber's work then every customer will find himself alone in the waiting room and will vainly open the door only to find the barber at work. This corresponds to the more obvious solution to the producer–consumer problem.

Algorithm 4.2

Barber As before, except that if the waiting room is empty then go to sleep on the bench in the waiting room.

Customer If there are other customers or if the waiting room is empty then wait your turn. If the barber is sleeping in the waiting room then wake him.

In Algorithm 4.2, a customer will wait without opening the door to no avail. Eventually the barber will finish with the previous customer and invite the new one in. Only if the barber is actually waiting (sleeping) will a customer have to wake him.

4.6 THE BOUNDED BUFFER

An infinite buffer is not realistic. There are two basic techniques to bound the size of a buffer. The first is the *circular buffer* where the index of the array b is computed modulo the finite size of the array. That is, the data is wrapped around from the end of the array to its start. The code for a circular buffer is shown in Fig. 4.12 where n is the size of the array b. See Fig. 4.13 to understand the boolean conditions that define empty and full buffers.

```
(* producer *)
produce;
wait until ( (in>=out) and (in−out<n)
             or (in<out) and (out−in>1) );
append;
if in=n then in := 1 else in := in+1;
(* consumer *)
wait until (in <> out);
take;
if out=n then out := 1 else out := out+1;
consume;
```

Fig. 4.12.

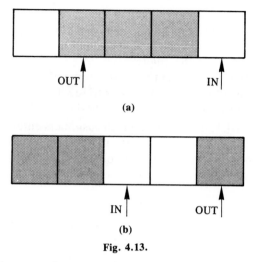

(a)

(b)

Fig. 4.13.

There is always at least one free space in b so that $in=out$ can be unambiguously identified as an empty buffer, not a full one.

The other method of bounding the size of buffers is to use two or more distinct buffers, usually of the same size. When one is filled by the producer it is passed on to the consumer which for its part promises to return the empty buffers to the producer for re-use. The circular buffer is very easy to program and very thrifty of space since there is only one buffer element of overhead. The multiple buffers need to be programmed with some care since a mutual exclusion mechanism must be invoked during the transfer of a buffer from one process to another. Multiple buffers can waste space. Suppose the producer has filled b_1 but the consumer still has a few elements left to consume in buffer b_2 (Fig. 4.14). Then the producer must wait for the consumer even though almost half of the overall buffer space is currently empty.

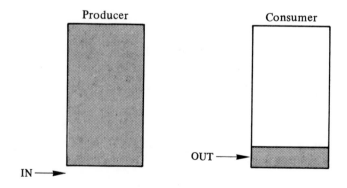

Fig. 4.14.

Multiple buffering is used in several situations. I/O equipment that uses direct memory access may not have the ability to work with a circular buffer. Typically such equipment accepts commands that consist of an address and a length of a memory segment into which the data is to be written. This translates nicely into a discrete buffer to be filled and passed to the program. Another important use for multiple buffers is in communication systems where the buffer requirements are highly variable in time. Rather than allocate a large circular buffer permanently to each terminal, it is better to construct a "pool" of many small buffers. An inactive terminal need not be assigned any buffers; active terminals can be assigned a large number of buffers from the pool to be returned once the data transfer is completed.

To change the producer–consumer problem to handle bounded buffers is very simple. Just as the general semaphore n counts the number of elements of the buffer currently filled by data, the program in Fig. 4.15 uses another semaphore e to count the number of empty spaces! When e reaches zero there are no empty spaces and the producer blocks until the consumer removes some of the data. The solution has a pleasantly symmetric form.

```
program    boundedbuffer;
const      sizeofbuffer = ... ;
var        s: (* binary *) semaphore;
           n: semaphore;
           e: semaphore;
procedure producer;
begin
  repeat
    produce;
    wait(e);
    wait(s);
    append;
    signal(s);
    signal(n)
  forever
end;
procedure consumer;
begin
  repeat
    wait(n);
    wait(s);
    take;
    signal(s);
    signal(e);
    consume
  forever
end;
```

```
begin (* main program *)
  s := 1;
  n := 0;
  e := sizeofbuffer;
cobegin
  produce; consume
coend
end.
```

Fig. 4.15.

4.7 EXERCISES

4.1 Write several tests until you thoroughly understand the difference between Figs. 4.9 and 4.10.

4.2 *Conway's Problem*: Write a program to read 80-column cards and write them as 125-character lines with the following changes. After every card image an extra blank is inserted. Every adjacent pair of asterisks ** is replaced by ∧.

Of course Conway's problem can be solved by a single sequential program. However, it is difficult to be sure that you have taken care of all of the special cases such as pairs of asterisks at the end of a card and so on.

The problem has an elegant solution as three concurrent processes. One process *read* reads the cards and passes characters through a one character buffer to a process *squash. read* also passes the extra blank at the end of every card image; *squash*, which knows nothing about 80-column cards, simply looks for double asterisks and passes a stream of modified characters to a process *print. print* takes the characters and prints them as 125-character lines.

4.3 Write a program to solve Conway's problem with the additional requirement that there be a 10-character buffer between each pair of processes: *read* and *squash*; *squash* and *print*. Use mutual exclusion on the bounded buffer as shown in Fig. 4.15.

4.4 In Fig. 4.12, simplify the condition in the producer's *wait until* clause.

4.5 Write a scenario that shows that *signal* must be a primitive instruction and not simply $s := s+1$.

4.6 Add a semaphore to **program** *increment* (Fig. 1.5) so that it always prints 40.

4.7 Write *sortprogram* so that *merge* is a third concurrent process.

4.8 Write a scenario showing lockout for the semaphore solution to mutual exclusion of three processes.

4.9 If you want to pass a semaphore as a parameter, should it be passed as a value parameter or a reference parameter (**var** in Pascal), or does it not make any difference? If you think that there is a difference, what happens if you make the wrong choice?

4.10 Why is the program in Fig. 4.16 not a solution to the problem of allowing at most k processes into a critical region (using binary semaphores only)?

(a) Show that if $k=2$ and $n=4$ it is possible to have $delay=2$ contrary to the requirement that $delay$ is a binary semaphore.

(b) *Show this even for $k=2$ and $n=3$.

```
program    mutualexclusion;
const      n = ... ;  (* number of processes *)
           k = ... ;  (* number in critical section *)
var        count: integer;
           s: (* binary *) semaphore;
           delay: (* binary *) semaphore;
procedure process (i: integer);
var    m: integer;
begin
  repeat
    wait(s);
    count := count - 1;
    m := count;
    signal(s);
    if m <= -1 then wait(delay);
    crit;
    wait(s);
    count := count + 1;
    if count <= 0 then signal(delay);
    signal(s)
  forever
end;
begin (* main program *)
  count := k;
  s := 1;
  delay := 0;
  cobegin
    process(1);
    process(2);
    . . .
    process(n)
  coend
end.
```

Fig. 4.16.

4.11 (Kessels and Martin) A split binary semaphore is a pair of binary semaphores x and y such that the formula $0 <= x+y <= 1$ is always true (i.e., is an invariant). Study the simulation (Fig. 4.17) of general semaphore s by the split binary semaphore.

A program using the general semaphore *s* should now declare *s* as an integer, assign it an initial non-negative value and then use the procedures *genwait(s)* and *gensignal(s)* below to simulate *wait* and *signal* for a general semaphore. The program must also declare and initialize global variables as shown.

```
count: integer; (* initially 0 *)
x: (* binary *) semaphore; (* initially 1 *)
y: (* binary *) semaphore; (* initially 0 *)
procedure genwait(var s: integer);
begin
  wait(x);
  while s <= 0 do
    begin
      count := count +1;
      signal(x); wait(y);
      count := count -1;
      if count=0 then signal(x) else signal(y);
      wait(x)
    end;
  s := s -1
  if count=0 then signal(x) else signal(y)
end;
procedure gensignal(var s: integer);
begin
  wait(x);
  s := s +1;
  if count=0 then signal(x) else signal(y)
end;
```

Fig. 4.17.

4.12 (Manna and Pnueli) Write a concurrent program to compute the binomial coefficient $(n\ k) = n(n-1) \ldots (n-k+1) / 1(2) \ldots (k)$, for $0 <= k <= n$. Let process P_1 multiply n then $n-1$ then $n-2$ and so on into a global variable x while process P_2 multiplies 1 and 2 and so on into a local variable y. Synchronize P_1 and P_2 so that P_2 executes $x := x$ **div** y when y, in fact, divides x. (*Hint i!* always divides $j(j+1) \ldots (j+i-1)$.)

4.13 *(Roussel) Write a concurrent program to test if two binary trees have the same leaves. There will be three processes: P_i, $i=1, 2$, will find the next leaf of tree i. When two "next" leaves have been found, P_3 will test them for equality. The trees can be declared in Pascal-S as follows:

```
const        maxnodes=40;
             leftson=0;
             rightson=1;
             nodevalue=2;
type         node=array[0 .. 2] of integer;
             treetype=array[ .. maxnodes] of node;
var          tree: array[1 .. 2] of treetype;
```

4.14 *Study the implementation kit and describe the changes that must be made so that the kit will catch misuse of a semaphore: tests and assignments (except in the main program). What changes must be made to differentiate general from binary semaphore?

4.15 (Parnas) Figure 4.18 is a solution to a problem called the Cigarette Smoker's problem. Each one of the three *agents* supplies two of three possible resources. There are three *smokers* each of which needs exactly the pair of resources supplied by one of the *agents*. Study (and test) the solution and answer the following questions.

(a) What is the function of the *helper* processes?
(b) Explain how termination of the tasks is accomplished. What is the purpose of $t := 0$ in **procedure** *forcetermination*?
(c) Even though the program terminates, there are certain bugs that may occur on the final execution of each process. Write examples for these bugs and a better program that terminates correctly.
(d) Do you see a problem that might occur with the semaphores $s[1]$, $s[2]$, $s[4]$? How can this problem be solved?

```
program cigarette;
const trips = 20;
var resource: array[1 .. 3] of (* binary *) semaphore;
    s: array[1 .. 6] of semaphore;
    mutex, sem: (* binary *) semaphore;
    t: integer;
    finished: array[1 .. 3] of boolean;
procedure forcetermination;
var i: integer;
begin
  t := 0;
  for i := 1 to 3 do signal(resource[i]);
  for i := 1 to 6 do signal(s[i])
end;
procedure agent(n, res1, res2: integer);
var i: integer;
begin
  for i := 1 to trips do
    begin
      wait(sem);
      signal(resource[res1]);
      signal(resource[res2]);
      writeln(' agent ' , n)
    end;
  finished[n] := true
end;
procedure helper(n, increment: integer);
begin
  repeat
    wait(resource[n]);
    wait(mutex);
    t := t + increment;
    signal(s[t]);
    signal(mutex)
```

```
        until finished[1] and finished[2] and finished[3];
        forcetermination
      end;
      procedure smoker(n, index: integer);
      begin
        repeat
          wait(s[index]);
          t := 0;
          writeln ('smoker', n);
          signal(sem)
        until finished[n]
      end;
      begin (* main program *)
        for t := 1 to 3 do resource[t] := 0;
        for t := 1 to 6 do s[t] := 0;
        for t := 1 to 3 do finished[i] := false;
        t := 0;
        mutex := 1;
        sem := 1;
        cobegin
          agent(1,2,3); helper(1,1); smoker(1,6);
          agent(2,1,3); helper(2,2); smoker(2,5);
          agent(3,1,2); helper(3,4); smoker(3,3)
        coend;
        writeln('smoking is dangerous')
      end.
```

Fig. 4.18.

5 MONITORS

5.1 INTRODUCTION

While the semaphore is an elegant low level synchronization primitive, an operating system built on semaphores alone is subject to disaster if even one occurrence of a semaphore operation is omitted or mistaken anywhere in the system. We should like to have a more structured synchronization tool. In sequential programming, one can write procedures out of simple statements and (hopefully) use the procedures by external specification only—by knowing the name and parameters of the procedure. The monitor is designed to allow concurrency while retaining the advantage of a structured construct.

The monitor grew out of two distinct ideas that are in common use. The first is the monolithic monitor used in current operating systems. Most of these systems are in effect single programs that centralize all critical functions. If a message must be passed from P_1 to P_2 then P_1 passes it to "big brother" monitor M with a request to forward it to P_2, or at least P_1 requests permission to pass the message. Similarly, resource allocation is centralized in M. Such monitors are supported by hardware facilities that ensure the privileged position of the monitor: M runs in an uninterruptable mode thus guaranteeing mutual exclusion; only M can access certain areas of memory; only M can execute certain instructions such as I/O instructions.

The monitors we study are decentralized versions of the monolithic monitor. Each monitor will be entrusted with a specific task and in turn it will have its own privileged data and instructions. Thus if M_1 is the only monitor that can access variable v_1 then we are ensured of mutual exclusion of access to v_1 because M_1 will be uninterruptable or as we say in the abstraction: entry to a monitor by one process excludes entry by any other process. In addition, since the only processing that can be done on v_1 is the processing programmed into the monitor, we are assured that no other assignments or tests are accidently made on v_1. We can design different monitors (or even different instances of the same monitor) for different tasks. Thus the system is both more efficient because execution of distinct monitors can be done concurrently, and more robust because a change in one monitor cannot surreptitiously change a variable in another monitor.

The other idea behind monitors is that of structuring data and structuring accesses to data in a programming language. Pascal was the first language specifically designed to structure data by *typing*. The purpose of data typing is to prevent indiscriminate mixing of data that have no purpose being mixed even though their representation may be identical. For example, even though it might occasionally be convenient to add a number to a character (and PL/I will be happy to do so), Pascal takes the view that such a statement is almost certainly in error and refuses to compile the statement. If this is what you really want then you must placate the compiler by writing an explicit conversion.

While typing is successful as far as it goes, it applies only to the data itself. There is no way in Pascal to define a type which consists of integers that may only be added or subtracted. There is no way of asking the compiler to flag a multiplication as unreasonable (in an accounting system, it might not make sense to multiply two sums of money—only to add and subtract them or to multiply a sum by an interest rate or a time period). The idea of typing by the operations performed is found in the language Simula 67. A *class* in Simula 67 is a data declaration together with a set of procedures which define the only legal operations that may be performed on the data. The Simula class has been combined with the Pascal data type in the Ada *package* feature which provides a carefully designed mechanism that allows the programmer to structure his program to reflect his knowledge of the properties of the data.

A *monitor* is a class that can be executed in turn by several processes. For example, even though the buffer is an array, there is no reasonable thing to do with a buffer except to append a new element or extract an old one. A random access to a buffer array is probably a mistake. To sort the elements of a buffer simply contradicts the definition of a buffer as an area of memory from which data is extracted in the same order that it was appended. In the monitor notation, once the buffer monitor has been written, there is no way to even express such a mistake. The compiler will refuse any access by a process to a buffer except through procedures *append* and *take*.

In this chapter we formally define the monitor as a programming primitive and give examples of its use. We show how to implement monitors on a system with semaphores. Monitors have been used very successfully in concurrent programming languages such as Concurrent Pascal and CSP/k.

5.2 DEFINITION OF MONITORS

A monitor is written as a set of (global) variable declarations followed by a set of procedures (which may be parameterized). The monitor has a body (**begin** . . . **end**) which is a sequence of statements that is executed immediately when the program is initiated. The body is used to give initial values to the monitor variables. Thereafter, the monitor exists only as a package of data and procedures.

The variables in the monitor are directly accessible only within the monitor procedures. Communication between a monitor and the outside world is through the parameters of the procedures. In the usual vocabulary of programming languages, the scope of the monitor variables is the monitor (= set of monitor procedures). Since the monitor is a static object: variable declarations and procedure declarations, the only way to execute the monitor is for a process to call a monitor procedure.

```
program producerconsumer;
const     sizeofbuffer = . . .;
  monitor boundedbuffer;
  b: array [0 . . sizeofbuffer] of integer;
  in, out: integer;
  n: integer;
  procedure append (v: integer);
  begin
    if n = sizeofbuffer+1 (* the buffer is full *)
      then "wait until not full",
    b[in] := v;
    in := in+1;
    if in = sizeofbuffer+1 then in := 0;
    n := n+1;
    "signal that the buffer is not empty"
  end;
  procedure take (var v: integer);
  begin
    if n=0 (* the buffer is empty *)
      then "wait until not empty";
    v := b[out];
    out := out+1;
    if out = sizeofbuffer+1 then out := 0;
    n := n - 1;
    "signal that the buffer is not full"
  end;
```

```
    begin (* monitor body *)
      in := 0;
      out := 0;
      n := 0
    end;
(* end of monitor boundedbuffer *)
procedure producer;
var        v: integer;
begin
  repeat
    produce (v);
    append (v)
  forever
end;
procedure consumer;
var        v: integer;
begin
  repeat
    take (v);
    consume (v)
  forever
end;
begin (* main program *)
  cobegin
    producer; consumer
  coend
end.
```

Fig. 5.1.

Figure 5.1 shows a solution to the producer–consumer problem using a bounded buffer. It consists of a program—*producerconsumer*—which contains a monitor—*boundedbuffer*—and two concurrent processes—*producer* and *consumer*. When the program is initiated, the monitor body is first executed. Then the main program is executed which causes the initiation of the concurrent processes. The monitor procedures *append* and *take* sit passively until called from a process.

If a producer wishes to append an element v to the buffer, all it need do is call the monitor procedure *append*(v). Similarly, the consumer can call *take*(v) to obtain the next element. We require that entry to a monitor be done under mutual exclusion. Thus either the producer is executing *append* or the consumer is executing *take* (or neither). Then the operations on the variable n by both processes do not interfere with each other.

This solution has certain advantages over the unstructured semaphore solution aside from the mere fact of the protection of the variables from outside interference and the structuring of the accesses. Since the mutual exclusion is automatically guaranteed by the monitor, there is no counterpart to a bug caused by omitting a *signal* to release mutual exclusion. If you forget the **end** statement of monitor procedure, that is a compilation error just as it would be in an ordinary program. In addition, the waiting and signalling is programmed within the monitor. The users of the monitor need only call a procedure. Thus once a monitor is correct, it will be correct for every instance used by every set of processes. In the case of the unstructured semaphore, the correctness depends upon semaphore operations that must be explicitly programmed into every process.

For synchronization we need some sort of wait–signal commands. The semaphore commands served two purposes. One is to provide a block–wakeup facility and the other to maintain a count. Since the counts can now be explicitly contained as integer variables in the protected monitor data, it is sufficient to provide a block–wakeup facility. Just as several semaphores may be needed in one program, so one monitor may need several wait–signal pairs. We define a new type of variable called a *condition* variable. If c is a condition variable then there are two commands that can be applied to c: *wait(c)* and *signal(c)* (these will now be defined for monitors; they are not to be confused with the commands of the same name for semaphores—alternatively call the semaphore commands P and V):

wait(c)   The calling process is blocked and is entered on a queue of processes blocked on this condition, i.e. have also executed *wait(c)* commands. Unlike semaphores we assume that the queues are FIFO.

signal(c)   If the queue for c is not empty then wake the first process on the queue.

Executing *signal(c)* when there are no processes waiting in the queue for c is a no-operation and leaves no traces (again unlike the semaphore). The commands may be worded: "I am waiting for c to occur" and "I am signalling that c has occurred". Of course it is the responsibility of the monitor programmer to ensure that c has occurred when the *signal* is issued.

The execution of *wait(c)* releases the mutual exclusion on the entry to the monitor. That this must be true is obvious. If P_1 is blocked waiting for c to occur, some process P_2 must do the signalling. If P_1 is not assumed to release the mutual exclusion then P_2 can never enter and the system is deadlocked. If we model the monitor as an igloo, the blocked processes can be considered to be in a deep-freezer in the basement. They do not take up room in the igloo so other processes can move around. Of course when a process signals, we must arrange either for the signalling process to leave the igloo immediately or for the blocked process to wait temporarily until the signalling process does leave. This ensures that there is only one process at a time in the igloo.

In the meantime we make the restriction that there be at most one *signal* per procedure and that it be the last statement in the procedure. Thus the signalling process leaves the monitor immediately after the signal. This evades the question raised in the last paragraph; we shall return to it in the last section.

Immediate Resumption Requirement Let a process execute *signal(c)* and suppose that there are processes in the queue for *c* as well as processes waiting to enter the monitor by a normal procedure call. Then the process on the head of the queue for *c* is the next process to enter the monitor; in particular it has priority over the processes which are trying to enter the monitor by procedure call.

Let us see why this requirement is needed. Suppose that process P_1 has noted that the condition is fulfilled and signals. If a process P_3 is allowed to enter the monitor (by a procedure call) before a process P_2 (which is waiting on condition *c*), then conceivably P_3 could cause *c* to become false. For example, if P_1 is a producer signalling buffer-not-empty and both P_2 and P_3 are consumers, then it would be fatal if the interloper P_3 consumed the single data element before the awakened process P_2 is allowed to proceed. Under the Immediate Resumption Requirement, however, P_2 can assume that whatever P_1 checked immediately before issuing the signal is still true because no interloper could falsify it between the signal and the resumption of the first blocked process.

Finally, to complete the bounded buffer example, we must declare two conditions in the variable declaration part of the monitor:

> *notempty, notfull: condition*;

and replace the phrases in quotes by the commands:

> *wait(notfull)*
> *signal(notempty)*
> *wait(notempty)*
> *signal(notfull)*,

respectively.

5.3 SIMULATION OF THE SEMAPHORE

The second example we describe is the implementation of a binary semaphore by a monitor. This is important theoretically since it shows that we have not lost any expressive power in the transition from semaphores to monitors. Practically, systems based on monitors are relatively common. If you have a ready-made semaphore algorithm you might want to run it as is on the monitor-based system.

The monitor (Fig. 5.2) will use a *boolean* variable *busy* which will indicate whether or not a *wait* operation has been completed on the semaphore (and hence that the critical section is "busy"). If the semaphore is busy, we must wait until it is not busy so according to the wording of the monitor operations we might as well call the condition *notbusy*. The two procedures will be called *P* and *V* to avoid confusion here with the monitor operations.

```
program    mutualexclusion;
   monitor  semaphoresimulation;
   var      busy: boolean;
            notbusy: condition;
   procedure P;
   begin
     if busy then wait(notbusy);
     busy := true
   end;
   procedure V;
   begin
     busy := false;
     signal(notbusy)
   end;
   begin (* monitor *)
     busy := false
   end;
procedure p₁;
begin
   repeat
     P;
     crit1;
     V;
     rem1
   forever
end;
procedure p₂;
begin
   repeat
     P;
     crit2;
     V;
     rem2
   forever
end;
```

begin (* *main program* *)
  cobegin
    p_1; p_2
  coend
end.

Fig. 5.2.

Let us check that the various requirements are fulfilled. There is at most one *signal* in every procedure and when it does occur, it is the last statement in *V*. If a *signal* is executed when there are blocked processes then it is executed when *busy* is false; hence, a blocked process which is awakened need not check that *busy* is now false and can proceed to set it true. Note how the mutual exclusion of the monitor entry prevents the bugs we once had when two processes simultaneously checked a variable such as *busy* and then set it.

This implementation of a semaphore is the strongest implementation possible because of the FIFO assumption on the queue of processes blocked on a condition. The semaphore definition does not require FIFO but it certainly does not forbid it as a scheduling strategy. Thus any semaphore algorithm proved correct under a weaker assumption is still correct.

A condition variable does not have a value in the usual sense of the word and hence no initialization is needed. To be more precise, every condition variable is implicitly initialized to the empty queue of processes blocked on it.

It seems that we need a different monitor for each semaphore. In practice, systems using monitors allow one to declare multiple instances and even parameterized sets of monitors. If the monitor procedures are written as re-entrant[†] procedures (Pascal procedures are automatically re-entrant), it is sufficient to allocate new instances of the global variables for each instance of a monitor. In the case of the semaphore monitor, this means a new boolean variable *busy* and a new queue for the condition *notbusy*. But that is exactly the amount of storage needed for a semaphore under the direct definition!

5.4 THE READERS AND WRITERS PROBLEM

A generalization of the mutual exclusion problem is the problem of the readers and writers. The prototype for the abstract problem is an on-line

---

† A procedure is called *reentrant* if its code can be used simultaneously by several processes. This is accomplished by writing pure code: code which is not self-modifying and which accesses all its data indirectly from an address usually kept in a register. Thus by merely changing the address in the register and remembering the current location of the program counter, the same code can be switched among several programs. In Pascal, procedures are always pure code and data is accessed by stack pointers. Hence by changing the stack pointers, a procedure can be forced to work on a different set of data.

transaction system such as a banking system not requiring mutual exclusion among several processes which only read the data. However, an update or any operation that writes data must be considered to be a critical section to avoid the type of bugs which should be thoroughly familiar by now. A monitor to solve this problem is given in Fig. 5.3.

```
program    readersandwriters;
  monitor  readwrite;
  var      readers: integer;
           writing: boolean;
           oktoread, oktowrite: condition;
  procedure startread;
  begin
    if writing or nonempty (oktowrite)
      then wait(oktoread);
    readers := readers+1;
    signal(oktoread)
  end;
  procedure endread;
  begin
    readers := readers-1;
    if readers = 0 then signal(oktowrite)
  end;
  procedure startwrite;
  begin
    if readers < > 0 or writing
      then wait(oktowrite);
    writing := true
  end;
  procedure endwrite;
  begin
    writing := false;
    if nonempty(oktoread)
      then signal(oktoread)
    else signal(oktowrite)
  end;
  begin (* monitor *)
    readers := 0;
    writing := false
  end;
```

```
procedure readprocess;
begin
  repeat
    startread;
    readthedata;
    endread
  forever
end;
procedure writeprocess;
begin
  repeat
    startwrite;
    writethedata;
    endwrite
  forever
end;
begin (* main program *)
  cobegin
    readprocess; readprocess; (* . . . *)
    writeprocess; writeprocess; (* . . . *)
  coend
end.
```

Fig. 5.3.

We have introduced a new primitive procedure into the monitor abstraction: *nonempty*(*condition*) which is a boolean valued function that returns true if and only if the queue for *condition* has blocked processes. If you examine the definition of *signal*(*condition*) you can see that it needs such an auxiliary function anyway so it is not unreasonable to require the implementer of monitors to provide the function for the programmer to use.

The number of readers that are reading or wish to read is counted by *readers*. Some process is writing if and only if *writing* is true. The two conditions are intended to mean literally what they say, e.g. "I am waiting for it being OK to read". We assume, of course, that the reading and the writing take a finite amount of time and that the processes do not accidentally terminate during reading and writing.

With the exception of the *signal* in *startread*, the monitor should not now be difficult to understand. The task of this *signal* is to perform what is known as a *cascaded wake-up*. If no readers are waiting when the *signal* is executed, then by definition it has no effect.

On the other hand, suppose that several readers $R_1, R_2, \ldots$ are blocked on *oktoread* because they began to execute *startread* during a write (and thus they found that the variable *writing* was true). When the write terminates, it will execute *endwrite* which signals *oktoread*. Since R_1 is at the head of the queue, it will be awakened and complete procedure *startread*. Obviously at this time, no one is writing and we might as well let in R_2 and all the other readers to execute concurrently with R_1. To accomplish this, R_1 executes *signal* (*oktoread*) which wakes R_2. But the same consideration applies to R_2 which proceeds to wake R_3. This cascade of signals thus wakes all the readers in the queue for *oktoread*. The final reader will execute a *signal* on an empty queue but of course this has no effect.

The problem of the readers and the writers leaves much room for imaginative variations on the question of priority. Consider the following possible sequence of requests: $R_1, R_2, W_1, R_3, \ldots$ Obviously R_1 and R_2 can read concurrently and W_1 has to wait for the termination of the reads, but what about R_3? It seems a pity to have R_3 wait both for the readers to finish and for the writer W_1 to finish. Thus we might let R_3 have priority over waiting writers (W_1) and let it read concurrently with R_1 and R_2. But now suppose that the scenario continues with a long sequence of readers: $R_4, R_5, \ldots$. If new readers arrive in rapid succession, W_1 will be indefinitely delayed. This is what we have called lockout.

In the current solution the rule is:

1. If there are waiting writers then a new reader is required to wait for the termination of a write.
2. If there are readers waiting for the termination of a write, they have priority over the next write.

The test on the condition *nonempty*(*oktowrite*) in *startread* ensures that if there are waiting writers (even if no-one is actually writing) then the new readers are blocked.

The **if** statement in *endwrite* ensures that if there are waiting readers at the termination of a write, they are given priority and awakened. The cascaded wakeup in *startread* will then wake up all currently waiting readers as required by the second clause of the rule. By the monitor requirement of immediate resumption, all the waiting readers will be awakened before any new readers are even allowed to enter the monitor. When new readers enter then they will block if there are waiting writers.

The selection of a priority scheme must be dictated by the application and no one scheme can be unreservedly recommended. In the exercises you can study programs which implement other schemes.

5.5 PROVING PROPERTIES OF MONITORS

Let us now give semi-formal proofs of some of the properties of this solution to the problem of the readers and the writers. Let R be the number of processes currently reading and let W be the number of processes currently writing.

When no process is currently executing a monitor procedure, the following formulae are true (i.e. the formulae are invariant outside the monitor).

(a) R = *readers*;
(b) $W > 0$ if and only if *writing* = *true*;
(c) *nonempty(oktoread)* only if (*writing* or *nonempty(oktowrite)*);
(d) *nonempty(oktowrite)* only if (*readers* $\neq$ 0 or *writing*);

Each of these statements is initially true and it must be checked (Exercise 5.7 (a)) that, if a statement is true upon entry into a monitor procedure, it is still true when the process exits the procedure.

Points to remember are: execution of a *wait* is also a way of exiting a monitor procedure; if a statement is true immediately before the execution of a *signal* then, by the immediate resumption of an awakened process, the truth of the statement is transfered to the resumed process.

The basic safety property required of a solution to the problem of the readers and the writers will be proven if we can show that the following formula I is invariant:

If $R > 0$ then $W=0$ and if $W > 0$ then ($W=1$ and $R=0$).

In words: if there are (active) readers then there are no writers, and if there are (active) writers then there is only one writer and no readers.

I is initially true since $R=W=0$. We show that I is always true by showing that any attempt to describe an execution sequence which falsifies I is unsuccessful.

1. Suppose $R > 0$ and $W=0$ (so that I is true) and then I is falsified by some process starting to write (so W will become 1).

By (a), $R > 0$ implies *readers* >0 so the process that wishes to write will *wait* in procedure *startwrite*. The only way this scenario could falsify I is if a *signal(oktowrite)* occurs. The *signal* in *endread* is executed only if *readers*=0, contrary to assumption. The *signal* in *endwrite* will also not be executed since there are no writers by the assumption $W=0$.

2. Suppose $R=0$ and $W > 0$ and then some process starts reading so that $R=1$, falsifying I.

$W>0$ implies *writing* = *true* by (b), so any process executing *startread* will *wait* on *oktoread*. Since $R=0$, there are no readers so *signal(oktoread)* is not executed in *endread*. Now I is assumed true so $W > 0$ implies $W=1$. Thus executing *signal(oktoread)* in *endwrite* upon termination of writing occurs when $W=0$ contradicting the assumption of this scenario.

3. $W=1$, $R=0$ and then some process starts writing to falsify the second clause of I.

This is impossible by the code in *startwrite*. The only signal possible is the one from *endwrite*, but then $R=W=0$ so I is not falsified.

The liveness properties of the solution are: if P wishes to read (or write) then eventually it will be allowed to do so. Let us prove the liveness of reading and leave the proof for writing as an exercise (see Exercise 5.7 (b)).

If P wishes to read then, it must successfully complete the execution of *startread*. If it cannot do so, it must become enqueued indefinitely on *oktoread*. We prove that *oktoread* is signalled indefinitly often. Since the queue is FIFO and the number of processes $P_1 \ldots, P_k$ ahead of P is finite, eventually P must be awakened and allowed to read.

By (c) above, either *writing=true* or *nonempty(oktowrite)*. If *writing = true* then by (b), $W>0$ so some process Q is writing.

  (i) (Q is writing) implies eventually (Q executes *endwrite*) by the assumption on critical sections that writing terminates.
 (ii) (Q is in *endwrite*) implies eventually (Q executes *signal(oktoread)*), by the assumption that P is indefinitely enqueued on *oktoread* so that *nonempty(oktoread)* is true.
(iii) (Q executes *signal(oktoread)*) implies eventually (P_1 executes *signal(oktoread)*), by the immediate resumption of a waiting process and the code in *startread*.
 (iv) (P_1 executes *signal(oktoread)*) implies eventually (P_2 executes *signal(oktoread)*), by immediate resumption and the code in *startread*.
  (v) (Q is writing) implies eventually (P_k executes *signal(oktoread)*), by (i)–(iv) and induction.

It easily follows that eventually P must successfully complete *startread* and commence reading.

Suppose now that *writing=false* but *nonempty(oktowrite)=true*. By (d) and the assumption that *writing=false, readers≠0* and thus by (a) there must be $P_1, \ldots, P_k$ currently reading.

  (i) ($P_1, \ldots, P_k$ are all the reading processes) implies eventually ((some P_i executes *endread*) and ($P_1, \ldots, P_{i-1}, P_{i+1}, \ldots, P_k$ are all the reading processes)), since reading terminates and *nonempty(oktowrite)* blocks new readers.
 (ii) ($P_1, \ldots, P_k$ are all the reading processes) implies eventually ((some P_j executes *endread*) and (there are no reading processes)), by induction.
(iii) ($P_1, \ldots, P_k$ are all the reading processes) implies eventually (P_j executes *signal(oktowrite)* and then *writing=true*), using immediate resumption and *nonempty(oktowrite)=true*.

This reduces to the previous case for which we have already shown liveness.

5.6 THE SIMULATION OF MONITORS BY SEMAPHORES

We now give an algorithm for transforming a program using monitors into a program that uses semaphores. This will show that monitors are no more powerful than semaphores and hence that the decision to use monitors can be made solely on the basis of their contribution to the clarity and reliability of the resulting system. This transformation is concerned only with the dynamic behavior of the concurrent system. The static protection of monitor variables should still be implemented, if possible, by a facility such as the Simula 67 class or the Ada package.

The mutual exclusion of the monitor procedures is easily simulated by a binary semaphore. There will be a semaphore s (initially 1) for each different monitor and each procedure of a monitor will commence with $wait(s)$ and terminate with $signal(s)$ just as we solved the critical section problem with semaphores. For each condition $cond$ we need an integer variable $condcount$ to count the number of waiting processes and a binary semaphore $condsem$ to actually block the waiting processes. The initial value of $condcount$ is 0. That of $condsem$ is also 0 because a process executing the monitor $wait$ always blocks. Each command $wait(cond)$ of the monitor is now coded:

$$condcount := condcount + 1;$$
$$signal(s);$$
$$wait(condsem);$$
$$condcount := condcount - 1;$$

The $signal$ on the semaphore s is to release the mutual exclusion on the entry to the monitor in order to allow other processes to enter, including (hopefully) one which will eventually signal.

Remember that we have restricted the $signal(cond)$ to be the last command of a procedure. Hence the release of a blocked process from the semaphore $condsem$ can be combined with the release of the mutual exclusion in the following code:

if $condcount > 0$ **then** $signal$ $(condsem)$ **else** $signal(s)$

This implements our restriction that, if there are processes waiting on a condition, they have priority over processes waiting at the monitor entry points. The blocked processes are waiting on $condsem$; the processes wishing to enter are waiting on s. So only if there are no blocked processes ($condcount = 0$) is the monitor entry freed. Note that the signalling process must have passed a $wait(s)$ semaphore upon entry. Thus the awakened process inherits the outstanding mutual exclusion from the signalling process. The debt is made good when the awakened process terminates and executes $signal(s)$. Of course it could avoid this by awakening still another process.

It is interesting that, according to our definition of semaphores, the value of the semaphore *condsem* is never 1. It is initially 0 and *signal(condsem)* is only executed if some process is waiting on *condsem*. Thus *condsem* is never incremented.

The only feature of the monitor that we cannot implement is the FIFO assumption on the queue because the semaphores are not FIFO.

In Fig. 5.4, we show how this translation can be carried out for the producer–consumer program in Fig. 5.1.

```
program    producerconsumer;
const      sizeofbuffer= . . . ,
var        b: array[0 . . sizeofbuffer] of integer;
           in, out: integer;
           n: integer;
           s: (* binary *) semaphore; (* for mutual exclusion *)
           notemptysem, notfullsem: (* binary *) semaphore;
           notemptycount, notfullcount: integer;
procedure append (v: integer);
begin
  wait(s);
  if n=sizeofbuffer+1 then
    begin
    notemptycount := notemptycount+1;
    signal(s);
    wait(notemptysem);
    notemptycount := notemptycount-1
    end;
  . . .
  if notfullcount > 0 then signal(notfullsem)
                      else signal(s)
end;
procedure take(var v: integer);
begin
  wait(s);
  if n=0 then
    begin
    notfullcount := notfullcount+1;
    signal(s)
    wait(notfullsem);
    notfullcount := notfullcount-1;
    end;
  . . .
  if notemptycount > 0 then signal(notemptysem)
                       else signal(s)
end;
```

procedure *producer*;
  . . .
procedure *consumer*;
  . . .
begin (* *main program* *)
  in := 0; *out* := 0; *n* := 0;
  s := 1;
  notemptycount := 0; *notfullcount* := 0;
  notemptysem := 0; *notfullsem* := 0;
  cobegin
    producer; *consumer*
  coend
end.

Fig. 5.4.

5.7   UNRESTRICTED SIGNALS

If we allow *signal*(*c*) to appear anywhere in a monitor procedure then we face the following dilemma. A primary advantage of the monitor is the immediate resumption of a signalled process while the condition is guaranteed. This saves needless looping and re-testing that is characteristic of semaphore algorithms. Thus we cannot let a signalling process continue after the signal because it might change the condition or execute another signal causing a third process to be immediately resumed.

On the other hand, the signalling process is not waiting for anything so it seems a pity to block it. Nevertheless, immediate resumption is so central to monitor programming that we retain it and instead block the signalling process. However, the signalling process is kept close at hand so that when the awakened process releases the monitor (by exiting or executing another *wait*) it has priority over other processes wishing to enter. Of course, nothing prevents the awakened process itself from signalling and thus joining the previous signaller.

In the igloo model (see Fig. 5.5), the igloo has only room for one process at a time. It has deep freezers in the basement for processes to wait on conditions. We add a closet with a spring loaded door.

Entering and exiting the igloo is on the basis of mutual exclusion—one at a time. If a process waits then it enters the freezer for the appropriate condition. If a process signals, it invites a frozen process to enter the igloo. The signalling process itself enters the closet so as not to interfere with the privacy of the awakened process. When the igloo is cleared, either because the process has exited or entered another freezer, the spring-loaded door flies open, thrusting the signaller into the igloo before a new process can enter. If awakened processes signal, they will be stacked in the closet on top

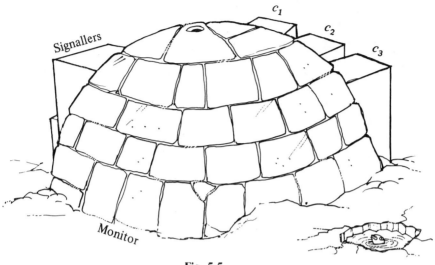

Fig. 5.5

of the previous processes.

We can now appreciate the simplification obtained by restricting the *signal* to being the last statement in a procedure. In the more general model, the signaller would emerge from the closet only to exit immediately. The signaller might just as well exit immediately after the signal and we need not build the closet into the igloo.

The simulation of the unrestricted monitor by semaphores is more complicated (Fig. 5.6). We need a new variable *urgent* and a new semaphore *usem* to count and block the signallers. In the simulation of *wait(cond)* and *exit*, the release of the monitor now checks *urgent* to see if there are waiting signallers. $urgent > 0$ will signify that there is some waiting signaller that is demanding to be released (by *signal(usem)*) before any new processes can be allowed to enter ·(by *signal(s)*).

In the simulation of *signal(cond)*, $condcount > 0$ signifies that there is in fact a process waiting for this condition. The process is awakened (by *signal(condsem)*) and then the signaller suspends itself (by *wait(usem)*).

```
entry: wait(s);
wait(cond): condcount := condcount+1;
    if urgent > 0                    (* if waiting signallers *)
        then signal(usem)            (* free a signaller *)
        else signal(s);              (* let a new process in *)
    wait(condsem);
    condcount := condcount−1;
```

```
signal(cond) : urgent := urgent+1;
    if condcount > 0 then
        begin                                (* if someone is waiting *)
        signal(condsem);                     (* let him in *)
        wait(usem)                           (* suspend yourself *)
        end;
    urgent := urgent−1;
    exit : if urgent > 0                     (* if a waiting signaller *)
        then signal(usem)                    (* free a signaller *)
        else signal(s);                      (* let in a new process *)
```

Fig. 5.6.

5.8 EXERCISES

5.1 Suppose that signals are not restricted but that some particular signal is in fact the last statement in its procedure. Code a simplification to simulation of monitors by semaphores for this case.

5.2 Show that even if the implementation of the semaphore is FIFO, the simulation of the monitor by semaphores is not.

5.3 Translate the monitor solution to the problem of the readers and the writers (Fig. 5.3) to a semaphore solution.

5.4 (Courtois *et al.*) Fig. 5.7 is a solution to the problem of the readers and the writers using semaphores. It is easy to see that r counts the number of readers, s guarantees mutual exclusion to variable r and *wsem* guarantees mutual exclusion to writing.

(a) Discuss the priority scheme of this solution.
(b) Suppose that you have a FIFO implementation of semaphores. How would that affect the answer?

```
program  readersandwriters;
var      r: integer;
         s, wsem: (* binary *) semaphore;
procedure readprocess;
begin
  repeat
    wait(s);
    r := r+1; if r=1 then wait(wsem);
    signal(s);
    readthedata;
    wait(s);
    r := r−1; if r=0 then signal(wsem);
    signal(s)
  forever
end;
```

```
procedure writeprocess;
begin
  repeat
    wait(wsem);
    writethedata;
    signal(wsem)
  forever
end;
begin (* main program *)
  r := 0;
  s := 1;
  wsem := 1;
  cobegin
    readprocess; (* . . . *)
    writeprocess; (* . . . *)
  coend
end.
```

Fig. 5.7.

5.5   (Courtois *et al.*) The solution in Fig. 5.8 uses a somewhat symmetrical code for both the readers and the writers except that in the writer, *wsem* is used to guarantee mutual exclusion in the execution of the write itself while, in the reader, *rsem* is used only by the writer to prevent readers from entering the critical section bracketed by s_1.

(a) Discuss the priority scheme of this solution.
(b) How would the solution change if the semaphores were FIFO?
(c) What would happen if s_3 were omitted? (*Hint* Show that s_3 ensures that $rsem=0$ implies that there is exactly one process waiting for *signal(rsem)*).

```
program    readersandwriters;
var        r,w: integer;
           s₁, s₂, s₃: (* binary *) semaphore;
           wsem, rsem: (* binary *) semaphore;
procedure readprocess;
begin
  repeat
    wait(s₃);
      wait(rsem);
    wait(s₁);
    r := r+1;
    if r=1 then wait(wsem);
    signal(s₁);
      signal(rsem);
    signal(s₃);
    readthedata;
    wait(s₁);
      r := r-1;
      if r=0 then signal(wsem);
    signal(s₁)
  forever
end;
```

```
procedure writeprocess;
begin
  repeat
    wait(s₂);
      w := w+1; if w=1 then wait(rsem);
    signal(s₂);
    wait(wsem);
    writethedata;
    signal(wsem);
    wait(s₂);
      w := w-1; if w=0 then signal(rsem);
    signal(s₂)
  forever
end;
begin (* main program *)
  r := 0; w := 0;
  s₁ := 1; s₂ := 1; s₃ := 1;
  rsem := 1; wsem := 1;
  cobegin
    readprocess; (* ... *)
    writeprocess; (* ... *)
  coend
end.
```

Fig. 5.8.

5.6 For each of the priority schemes (Section 5.4 and Exercises 5.4 and 5.5) try to think of an application in which that scheme is the reasonable one to require.

5.7 Complete the semi-formal proof of the solution to the problem of the readers and the writers.

(a) Prove the invariants (a)–(d).
(b) Prove the liveness of writing.
(c) *Prove the following statement S which expresses the priority scheme claimed for the solution.

S: If (Q enters *startwrite* before P enters *startread*) then (Q writes before P reads).

Hint How to prove precedences:

Let p and q be any two properties of programs. Then (p before q) is equivalent to (p is now true) or (q is now false and after executing a step of the program then (p before q)). Similarly, the negation of (p before q) is equivalent to (p is now false) and (q is now true or after a step then (p before q) is false).

Show that the negation of S is inductive, that is, if S is assumed false then after executing a step of the program S must still be false. Deduce that (Q enters *startwrite*) but never (Q writes) thus contradicting the liveness of writing shown in (b).

6 THE ADA RENDEZVOUS

6.1 INTRODUCTION

The semaphore and the monitor are centralized facilities. A process executing a monitor procedure has access to the single copy of the monitor variables and thus there must be an arbiter to enforce mutual exclusion. While these requirements are natural in a single computer (or any common memory system), they are difficult to implement in a distributed system. By a distributed system we mean a set of totally independent computers whose only connection is by sending and receiving messages. There may be no synchronization between sending and receiving a message. Messages may "pass" each other in transit. What we need is some sort of self-enforcing protocol. A process will decide to wait on its own initiative as in the busy wait algorithms of Chapter 3.

In those algorithms, each process agrees to enter a **while** loop if it needs to block itself. Each process decides by itself when to leave. This contrasts with the semaphore and the monitor signals which are required to wake another process.

Lamport's bakery algorithms discussed in the exercises of Chapter 3 are designed for distributed systems. Each process writes only into a single local variable. The test $n_1 > n_2$ in process P_1 can be interpreted as follows. Send a request to process P_2 to read the value of n_2. Wait until P_2 replies. Compare the received value with the current value of the local variable n_1.

Each process needs to be augmented with polling statements that periodically search for requests for data values from other processes. Alternatively this can be done by interrupts. The receipt of a message from the communication line will trigger an interrupt in the receiving process. This process will then identify the message and route it for appropriate action.

If we examine the protocol suggested in the last paragraph we find that the essential idea is the transfer of information at a predetermined point in each process. We call this a *rendezvous* of the two processes. The essence of any rendezvous is that the party which arrives earlier is required to wait. The alternative to a rendezvous is a buffered message system but aside from questions as to the size and number of buffers, we now have the question of who owns the buffers.

In classical systems the buffers belong to the "system", but in a distributed system, we do not want to single out any process as the boss. If that is what we want then the monitor formalism is sufficient. Since message passing systems are an obvious task for concurrent programming, we prefer to base the programs on an independent primitive such as a rendezvous.

The rendezvous was suggested by Hoare in a paper which is entitled "Communicating sequential processes" to contrast with Dijkstra's "Cooperating sequential processes" on which Chapters 3 and 4 are based. Instead of presenting Hoare's original work we choose to discuss the version that is used in the Ada programming language. Ada was designed specifically for real-time systems programming which inherently uses concurrency. The Ada facilities will probably become the dominant style for concurrent programming.

The examples will be written as executable Ada programs. However, we will try to use only Pascal-like features when possible and we will note the essential differences where necessary. This chapter is not a tutorial on Ada nor even on the full range of concurrent programming facilities in Ada. It is intended as an introduction to the rendezvous concept as used in Ada. Ada programmers can regard the previous chapters as a description of the scientific climate under which the Ada concurrent programming primitives were developed.

6.2 THE "ACCEPT" STATEMENT

The monitor has no life of its own. It is simply a collection of data and procedures that sit waiting to be invoked. In our model, the monitor is an igloo which is accessible to all processes that need it but which is just a building that does nothing on its own. Let us now imagine a story that will give us a feel for the rendezvous.

Several processes, P_i, are riding around in the snow on their dog sleds (Fig. 6.1). Process Q owns a lodge "Ada's Place" which happens to be strategically placed at an intersection of all the tracks followed by the P_i. Periodically, the P_i reach the lodge and wish to enter for a snack. If P_1 arrives before Q has arrived to open the lodge then P_1 crawls into his sleeping bag to wait for Q. When Q arrives they open the lodge, go inside and P_1 exchanges the fresh meat he has bagged for ready-to-eat meat sandwiches that Q prepares.

Unfortunately, Q's investment capital was too small and the lodge can only hold two processes and one load of meat and bread. Thus, if several

Fig. 6.1.

processes arrive at the same time or if several processes arrive when a transaction is in progress, then only one process P_1 can be in the lodge with Q. The others must wait until P_1 leaves the lodge to resume his hunt and Q leaves the lodge to ride his sled to the Bakery to replenish his supply of bread. When Q returns he can complete a similar transaction with the other processes.

Suppose now that a sudden blizzard has delayed the arrival of the P_i. Then Q must wait. Since it seems pointless to waste energy by heating the lodge when there are no customers, the lodge stays locked and Q keeps warm in a sleeping bag outside.

The essential features to be abstracted from this story are:

(i) the symmetrical waiting of the rendezvous—whoever arrives first is required to wait;
(ii) the completion of a single transaction during the rendezvous;
(iii) the mutual exclusion of the transaction conducted in the confinements of the lodge; and
(iv) the two-way exchange of information that is the method of communication between the processes in the rendezvous. Let us now examine the Ada program for this story (Fig. 6.2).

The **accept** statement has the syntax of a local procedure (without local data declarations) which can appear embedded within executable statements. The call to the **accept** statement is syntactically identical to a procedure call. This is done intentionally so that an operation can be implemented as a sequential procedure or a concurrent program without informing the user. However, unlike a monitor procedure, the **accept** statement belongs to process Q. Q must accept something in order to continue execution. As described in the story, if some P_i executes a call before Q executes the **accept** then P_i is blocked pending the rendezvous. Conversely, if Q executes the **accept** and there are no waiting processes P_i then Q is blocked pending the rendezvous. Once Q and some P_i have executed the pair **accept** *lodge* and *lodge* then we say that the rendezvous has occurred.

```
procedure Adalodge is
   numberofprocesses: constant := . . . ;
task type process;
task Q is
   entry lodge(meat: in food; sandwiches: out food);
end Q;
task body process is
   walrus, victuals: food;
begin
   loop
      walrus := huntwalrus;
      lodge(walrus, victuals);
      eat(victuals);
   end loop;
end process;
task body Q is
   bread: food;
begin
   loop
      bread := visitbakery;
      accept lodge(meat: in food; sandwiches: out food) do
         cook(meat);
         sandwiches := meat+bread;
      end lodge;
   end loop;
end Q;

   P: array(1 . . numberofprocesses) of process;
begin
   null;
end Adalodge;
```

Fig. 6.2.

Remark To run this program, you will have to supply the value of *numberofprocesses*; a representation for *food*; subprograms for *huntwalrus*, *eat, visitbakery, cook,* and finally, the **loop** should be terminating and some trace should be printed. (End of *remark.*)

The rendezvous is considered to be in force during the execution of the statements between the **do** of the **accept** statement and the corresponding **end**. In particular, the calling process is blocked for the duration of the rendezvous to prevent it from changing the values of the parameters until the exchange of information is complete. Thus the body of the **accept** acts like a critical section. Once the **accept** statement has terminated, the rendezvous has been completed and a fresh **accept** statement must be issued—during the next cycle of the loop—to effect another rendezvous (with the same process or with another). Of the two parameters, one is used to pass a value from the calling process and the other to return a value to the calling process. This is the two-way exchange of information between the processes.

That the rendezvous is suitable for distributed systems can be seen by the following sketch of how it might be implemented. P_i executes *lodge* by sending a signal to Q that it requests a rendezvous. P_i then suspends itself pending a reply from Q. The processor running Q registers the signal from P_i by an interrupt or by polling. Q eventually executes the **accept** statement and notes that P_i has registered a signal. Q replies to P_i that rendezvous has occurred and remains blocked until P_i acknowledges the reply with a message containing the parameters of the call. Q receives the parameters, executes the statements following the **do** and then returns the result parameters. Upon receiving the results, P_i can unblock and continue computation. The program can even be implemented in a distributed manner because once the number and type of the parameters are agreed upon, the processes can be independently designed and programmed.

Let us now see how the binary semaphore can be simulated by a rendezvous (Fig. 6.3). Here it is important to note that while the mutual exclusion problem for two processes was solved by invoking passive semaphore procedures, with the rendezvous we need to create a new semaphore process to mediate between P_1 and P_2. On the other hand, P_1 and P_2 no longer need to access the same variable or queue.

```
procedure mutualexclusion is
task semaphore is
   entry wait;
   entry signal;
end semaphore;
task body semaphore is
begin
   loop
      accept wait;
      accept signal;
   end loop;
end semaphore;
```

```
task P₁;
task body P₁ is
begin
  loop
    rem1;
    wait;
    crit1;
    signal;
  end loop;
end P₁;

task P₂;
task body P₂ is
begin
  loop
    rem2;
    wait;
    crit2;
    signal;
  end loop;
end P₂;
begin
  null;
end mutualexclusion;
```

Fig. 6.3.

When a process P_1 executes the call to *wait*, it must block until the semaphore process executes its **accept** statement and the rendezvous is achieved. There are no parameters to be passed and no statements to be executed within the critical section of the **accept** statement. Once P_1 has terminated the rendezvous, it is free to enter its critical section.

P_2 however will block when it tries to call *wait* because the semaphore process is waiting for a rendezvous with a *signal* call. Thus until some process (i.e. P_1) executes a call to *signal*, the semaphore process is blocked and hence so is P_2. When P_1 completes its critical section, it accomplishes a rendezvous with the semaphore process at **accept** *signal*. Then the semaphore process can commence its next cycle and accomplish a rendezvous with P_2.

Since the queues for the **accept** statements are required to be implemented as FIFO queues, this implements a FIFO semaphore: even if P_1 overtakes P_2 after completing the critical section, it will be placed after P_2 on the queue of the same **accept** *wait* statement. Note that the queues are FIFO in terms of time of arrival at the processor executing the **accept** statement. Thus, in a physically distributed system, closer processes may be able to overtake more remote ones. Lockout is not a problem, however, because once a process does enter the queue, it need only wait for the finite number of processes ahead of it to complete.

We have shown how to implement a single semaphore. If the system is to have several semaphores we will need multiple copies of the semaphore process so that calls to *wait* and *signal* can be parameterized to indicate which semaphore process to call. In Ada a task type can be defined and multiple instances created by ordinary variable declarations. Thus an array of semaphore tasks can be declared and accessed by a simple array index.

Ada language notes

1. Ada uses a "main" **procedure** where Pascal uses **program**.
2. The token **var** is not used before variable declarations.
3. **loop** . . . **end loop** is our **repeat** . . . **forever**.
4. Processes are called **tasks**. A **task** must have a specification part which declares the **entry**s: the name of the rendezvous, if any. The **body** is defined separately and contains the data and code local to the **task**.
5. An **in** parameter is read-only. For simple parameters this can be implemented by call-by-value: the actual parameter is evaluated and a copy of the value is passed to the procedure, or in this case the body of the **accept** statement. An **out** parameter is write-only. It can be used to copy a value from the procedure to a variable in the calling program.
6. To declare several identical tasks, we declare a **task type**. Then multiple instances may be declared in an **array**, as we have done.
7. The declaration of a task automatically initiates it. Thus the main program body is **null**. No **cobegin** . . . **coend** is necessary, since the token **task** calls for concurrent execution.

6.3   THE "SELECT" STATEMENT

Turning to the bounded buffer problem, we find that the **accept** statement is insufficient. We might attempt a solution with the program fragments of Fig. 6.4, using a buffer process between the producer and the consumer.

```
Producer:
  loop
    append(v);
  end loop;
Consumer:
  loop
    take(v);
  end loop;
```

Buffer:
```
loop
  accept append(v: in integer);
  accept take(v: out integer);
end loop;
```

Fig. 6.4.

The only execution sequence possible is *append, take, append, take*. This is the same as no buffer at all and we could just as well make the rendezvous directly between the producer and the consumer.

Producer:
```
loop append(v); end loop;
```
Consumer:
```
loop
  accept append(v: in integer);
end loop;
```

If you examine the bounded buffer solutions by semaphores and monitors you find that what is needed is some way of conditionally achieving a rendezvous. If the buffer is full, the rendezvous must only be with the consumer; if empty, only with the producer. If the buffer is neither full nor empty then the rendezvous can be with whichever of the processes is currently waiting for the rendezvous. If both are waiting then we do not really care with whom the rendezvous is made as long as the buffer process is not unfair. With this background it should be possible to follow the Ada solution to the bounded buffer (Fig. 6.5). The program shows just the buffer task; the producer and consumer are straightforward loops.

```
task boundedbuffer is
  entry append(v: in integer);
  entry take(v: out integer);
end boundedbuffer;
task body boundedbuffer is
  size: constant := ... ;
  b: array(0 .. size) of integer;
  inptr, outptr: integer;
  n: integer;
begin
  n := 0; inptr := 0; outptr := 0;
  loop
  select
    when n <= size =>
      accept append(v: in integer) do
        b(inptr) := v;
      end append;
      n := n+1;
      inptr := (inptr+1) mod size;
```

or
    when $n > 0 =>$
      accept *take*(v: **out** *integer*) **do**
        $v := b(outptr)$;
        end *take*;
        $n := n-1$;
        $outptr := (outptr+1)$ **mod** *size*;
    end select;
    end loop;
end *boundedbuffer*;

Fig. 6.5.

Before we discuss the new features of the **select** command, note that the critical section of the **accept** command does not encompass the entire buffer processing but only the physical exchange of data. The updating of the internal pointers need not block the producer nor the consumer who have no access to these local variables.

The **select** statement allows one to select between several alternatives separated by **or**. The alternatives are prefixed by **when**-clauses called *guards*. The guards are boolean expressions which establish what conditions must be true for an alternative to be a candidate for execution.

The execution of a **select** statement begins by evaluating all the guards. Then one of the *open* alternatives—alternatives with true guards—is selected for execution. In the bounded buffer if one of the guards is not true then the other must be ($n <= 0$ implies $n <= size$ and conversely, $n > size$ implies $n > 0$) so there is always an open alternative. If the buffer is empty, only the first alternative can be selected. This is the alternative that receives data from the producer. Similarly if the buffer is full, only the second alternative is open to allow the consumer to remove data from the buffer. In either case, of course, the buffer process will block waiting for the relevant rendezvous. Nothing is lost by not rechecking a closed alternative since the only way to empty a full buffer is by consuming.

The difference between the **select** statement and an **if** statement is seen in the case that both guards are open. Then if both the consumer and the producer are waiting for a rendezvous, we don't care which rendezvous is accomplished. An **if** statement must specify which statement is to be executed in this case.

But the **select** statement is even smarter than that. Suppose that the buffer is neither empty nor full, but that only the consumer is waiting for a rendezvous. If a choice is made between both open alternatives then we could blunder into blocking on an **accept** *append* statement for which no producer process is waiting. Thus if both alternatives are open but only one **accept** statement has a process blocked on it waiting for a rendezvous, the **select** will choose to execute the alternative that leads to an immediate rendezvous.

There is another possibility, namely that both alternatives are open but that neither process is ready. Rather than endlessly checking the guards (whose truth will not change) or arbitrarily blocking on one of the **accept** statements, the **select** statement will block simultaneously on both **accept** statements and execute the first rendezvous to be accomplished.

This is the essence of the "guarded commands" style of programming: avoid over-specification (as in an **if** statement) by allowing the computer as much freedom of choice as possible consistent with the correctness requirements of the program.

We now give a more formal description of the general **select** statement.

select
> **when** *condition*1 => **accept** *entry*1 **do** *statements* **end**;
> > *other statements*

or
> **when** *condition*2 => **accept** *entry*2 **do** *statements* **end**;
> > *other statements*

> . . .

else *statements*
end select;

Remark 1   The **else** clause is optional (see below).

Remark 2   A guard may be identically *true* in which case the **when** *true* => can be omitted.

Semantics of the **select** statement:
1. Evaluate all the guards to determine which alternatives are open.
2. If there are open alternatives, determine which **accept** statements in open alternatives have processes currently waiting for rendezvous.
3. If there are such processes, execute one of these alternatives. If there are several open alternatives with processes waiting for rendezvous, the selection among them is done arbitrarily.
4. If there are no open alternatives or no waiting processes, execute the **else** clause if there is one.
5. If there are no waiting processes and no **else** clause, wait for the first process to attempt a rendezvous with an **accept** clause in one of the open alternatives.
6. In the absence of an **else** clause, it is an error for there to be no open alternative.

We can demonstrate the general **select** statement by assuming in our story that the owner of "Ada's Place" has independent suppliers of both bread and meat and that his only task is to make sandwiches. Then if there are no processes waiting at the lodge, he can profitably use the time to prepare sandwiches (Fig. 6.6).

The guards are exhaustive (at all times at least one is true) and disjoint (no two are ever simultaneously true) so exactly one is open. The **else** clause is used to do some useful work if the second **accept** statement is open but does not have a process waiting for the rendezvous. If the P_i are fast eaters, they will have to wait but if they are often held up in blizzards, they can count on fast service when they do return to the lodge.

```
procedure Adalodge is
task type process;
task hunter;
task bakery;
task Q is
   entry delivermeat(mt: in food);
   entry deliverbread(br: in food);
   entry lodge(snack: out food);
end Q;

task body process is
   victuals: food;
begin
   loop
      explore;
      lodge(victuals);
      eat(victuals);
   end loop;
end process;

task body hunter is
   walrus: food;
begin
   loop
      hunt(walrus);
      delivermeat(walrus);
   end loop;
end hunter;

task body bakery is
   rolls: food;
begin
   loop
      bake(rolls);
      deliverbread(rolls);
   end loop;
end bakery;

task body Q is
   bread, meat, sandwiches: food;
```

```
procedure makesandwiches is
begin
    cook(meat);
    sandwiches := bread+meat;
    bread := 0;
    meat := 0;
  end makesandwiches;
begin
  bread := 0;
  meat := 0;
  sandwiches := 0;
  loop
  select
    when bread = 0 =>
      accept deliverbread(br: in food) do
        bread := br
      end deliverbread;
  or
    when meat = 0 =>
      accept delivermeat(mt: in food) do
        meat := mt;
        end delivermeat;
  or
    when ((bread <> 0) and (meat <> 0)) or
      (sandwiches <> 0) =>
        accept lodge(snack: out food) do
          if sandwiches=0 then
            makesandwiches; end if;
          snack := sandwiches;
          sandwiches := 0;
          end lodge;

        else
          if (bread <> 0) and (meat <> 0)
            and (sandwiches=0)
            then makesandwiches; end if;
        end select;
        end loop;
      end Q;

      P: array(1 .. numberofprocesses) of process;
      begin
        null;
      end Adalodge;
```

Fig. 6.6.

6.4   PROVING PROPERTIES OF THE RENDEZVOUS

The development of a satisfactory proof theory for synchronziation by communication is an area of current research. The concepts of invariants and eventuality propositions that have been so successful in proving synchronization by cooperation may need modification when applied to the rendezvous.

An important goal is to try to make the proofs distributed: the proof of one process ought not to require knowledge of the internal details of another process. It ought to be possible to construct communicating proofs: "If you promise to send me a message M_1 (I do not care how or why you do so) then I promise to send you a message M_2."

Let us look at the bounded buffer task of Fig. 6.5. First note that, since the guards are exhaustive, there is no deadlock. Since the execution of one alternative makes true the guard of the other alternative, there is no lockout if we assume fair selection among open alternatives. Another way of showing eventuality is to note that if the producer produces indefinitely without the consumer consuming, the buffer must fill and the producer's guard is falsified. The proof of liveness can thus be done by reasoning internal to the buffer task.

The safety property is that the elements must be consumed in the same order that they are produced. The internal behavior of the buffer task is very simple; what must be shown is that a value that is produced must become the newest element of the buffer and that the oldest element of the buffer is that which is consumed.

Suppose that the producer executes *append(v)*. Upon achieving the rendezvous, we can conclude that the value of the actual parameter has been communicated to the formal parameter *v* of the **accept** statement. Upon completion of the rendezvous (completion of the **accept** statement), this value has been transferred to *b[inptr]* and has become the newest element of the buffer.

The general rule is the transfer of information at the point of the rendezvous. For an **in** parameter *v*, whatever was true of *v* in the calling process before the rendezvous is true within the **accept** statement upon achieving rendezvous. For an **out** parameter (such as the *v* in *take*), whatever is true of *v* upon completion of the **accept** statement (such as that *v* is the oldest buffer element) is true of the actual parameter in the calling task.

6.5   EXERCISES

6.1   Solve Conway's problem in Ada.

6.2   Write programs for the reader–writer algorithms we have given.

6.3   Simulate a general semaphore in Ada.

6.4   Simulate a rendezvous by semaphores or monitors.

6.5   In Fig. 6.7 are two possibilities for a monitor-like *wait/signal* facility. Specify the behavior of these solutions. Which is more like a monitor?

6.6   Discuss the priority scheme of the solutions to the problem of the readers and the writers shown in Fig. 6.8 and compare them with the solutions in Chapter 5.

6.7   Prove the correctness of the programs in Figs. 6.2 and 6.6.

6.8   (Dijkstra) Write a program for partition by communication. Two disjoint sets of numbers S and T are given. If s and t are the number of elements of S and T, respectively, then upon completion of the program, S should contain the s smallest numbers in $S \cup T$ and T should contain the t largest numbers in $S \cup T$. A solution using two processes is outlined in Fig. 6.9. Prove its correctness. *Note* The termination of the task *upper* is not specified. If you program this solution in Ada, either learn about the **terminate** option in Ada or introduce explicit signals for termination.

```
        task monitorfacility1 is
           entry wait;
           entry signal;
        end monitorfacility1;
        task body monitorfacility1 is
           received: boolean;
        begin
           received := false;
           loop
           select
      accept signal;
      received := true;
   or
      when received =>
         accept wait;
         received := false;
   end select;
   end loop;
end monitorfacility1;

task monitorfacility2 is
   entry wait;
   entry signal;
end monitorfacility2;
task body monitorfacility2 is
begin
   loop
     accept signal;
     select
        accept wait;
     else null;
     end select;
   end loop;
end monitorfacility2;
```

Fig. 6.7.

```
task readersandwriters1 is
  entry startread;
  entry endread;
  entry startwrite;
  entry endwrite;
end readersandwriters1;
task body readersandwriters1; is
  readers: integer;
begin
  readers := 0;
  loop
  select
    accept startread,              - ·
    readers := readers+1;
  or
    accept endread;
    readers := readers−1;
  or
    when readers = 0 =>
      accept startwrite;
    accept endwrite;
  end select;
  end loop;
end readersandwriters1;


task readersandwriters2 is
  entry startread;
  entry endread;
  entry startwrite;
  entry endwrite;
end readersandwriter2;
task body readersandwriter2 is
  readers: integer;
begin
  readers := 0;
  loop
  select
    when startwrite'count = 0 =>
      accept startread;
    readers := readers+1;
  or
    accept endread;
    readers := readers−1;
  or
    when readers = 0 =>
      accept startwrite;
    accept endwrite;
    loop
      select
        accept startread;
        readers := readers+1;
      else goto l;
      end select;
```

```
    end loop;
⟪l⟫ end select;
    end loop;
end readersandwriters2;
```

Fig. 6.8.

Ada Language Notes

1. *startwrite'count* is a predefined function (called an attribute in Ada) that returns the number of tasks currently waiting to rendezvous with the **entry** *startwrite*.
2. ⟪l⟫ is a statement label that is the target of the corresponding **goto**.

```
task lower is
  entry sendmax(xx: in integer);
end lower;
task upper is
  entry sendmin(yy: in integer);
end upper;

task body lower is
  x, mx: integer;
begin
  mx := the maximum value in s;
  loop
    sendmax(mx);
    remove mx from S;
    accept sendmin(xx: in integer) do
      x := xx;
      end sendmin;
    add x to S;
    mx := the maximum value in S;
    exit when x=mx;
  end loop;
end lower;

task body upper is
  mn: integer;
begin
  loop
    accept sendmax(yy: in integer) do
      add yy to T;
      end sendmax;
    mn := the minimum value in T;
    sendmin(mn);
    remove mn from T;
  end loop;
end upper;
```

Fig. 6.9.

7 THE DINING PHILOSOPHERS

7.1 INTRODUCTION

The problem of the dining philosophers (posed by Dijkstra) is of great importance in concurrent programming research. The problem allows all of the pitfalls of concurrent programming to be demonstrated in a vividly graphical situation. It is a challenge to proposers of new primitives for concurrent programming. As a test of your understanding of concurrent programming principles we present several solutions using the various primitives we have learned.

The problem is set in a monastery whose five monks are dedicated philosophers. Each philosopher would be happy to engage only in thinking were it not occasionally necessary to eat. Thus the life of a philosopher is an endless cycle: **repeat** *think*; *eat* **forever**.

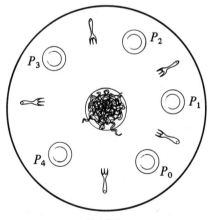

Fig. 7.1.

The communal dining arrangement is shown in Fig. 7.1. In the center of the table is a bowl of spaghetti that is endlessly replenished; there are five plates and five forks. A philosopher wishing to eat enters the dining room, takes a seat, eats and then returns to his cell to think. However, the spaghetti is so hopelessly entangled that two forks are needed simultaneously in order to eat. (It has been pointed out that this requirement is rather forced and that it would be more natural to set the problem in the Orient with a bowl of rice and chopsticks.)

The problem is to devise a ritual (protocol) that will allow the philosophers to eat. Each philosopher may use only the two forks adjacent to his plate. The protocol must satisfy the usual requirements: mutual exclusion (no two philosophers try to use the same fork simultaneously) and freedom from deadlock and lockout (absence of starvation—literally!) An additional safety property is that if a philosopher is eating then he actually has two forks.

7.2 FIRST ATTEMPT

```
program    diningphilosophers;
var        fork: array [0 . . 4] of (* binary *) semaphore;
           i: integer;
procedure philosopher(i: integer);
begin
  repeat
    think;
    wait( fork[i] );
    wait( fork[(i+1) mod 5] );
    eat;
    signal( fork[i] );
    signal( fork[(i+1) mod 5] )
  forever
end;
begin (* main program *)
  for i := 0 to 4 do fork[i] := 1;
  cobegin
    philosopher(0);
    philosopher(1);
    philosopher(2);
    philosopher(3);
    philosopher(4)
  coend
end.
```

Fig. 7.2.

The idea of the program in Fig. 7.2 is very simple. The binary semaphores ensure mutual exclusion in accessing the forks. The safety property is satisfied because eating is done only after two fork-semaphores have been successfully completed.

Unfortunately the solution deadlocks. Under perfect sychronization if the philosophers enter the protocol simultaneously and take the left forks then the state of the program is that all forks are 0 and all the philosophers are trying to complete *wait(fork[i+1])*. Since there is no process that can *signal*, the program is deadlocked.

7.3 SECOND ATTEMPT

```
program      diningphilosophers;
  monitor    forkmonitor;
  var        fork: array[0 . . 4] of integer;
             oktoeat: array[0 . . 4] of condition;
             i: integer;
  procedure takefork(i: integer);
  begin
    if fork[i] <> 2 then wait(oktoeat[i]);
    fork[(i+1) mod 5] := fork[(i+1) mod 5]−1;
    fork[(i−1) mod 5] := fork[(i−1) mod 5]−1
  end;
  procedure releasefork(i: integer);
  begin
    fork[(i+1) mod 5] := fork[(i+1) mod 5]+1;
    fork[(i−1) mod 5] := fork[(i−1) mod 5]+1;
    if fork[(i+1) mod 5]=2 then signal(oktoeat[(i+1) mod 5]);
    if fork[(i−1) mod 5]=2 then signal(oktoeat[(i−1) mod 5])
  end;
  begin (* monitor *)
    for i := 0 to 4 do fork[i] := 2
  end;
procedure philosopher(i: integer);
begin
  repeat
    think;
    takefork(i);
    eat;
    releasefork(i)
  forever
end;
```

```
begin (* main program *)
  cobegin
    philosopher(0);
    philosopher(1);
    philosopher(2);
    philosopher(3);
    philosopher(4)
  coend
end.
```

Fig. 7.3.

The previous solution failed because a philosopher is allowed to blindly take a free fork. In the present solution (Fig. 7.3), the array *fork* is now intended to represent the number of free forks available to a philosopher. Only if both forks are available will the philosopher take them (by completing procedure *takefork*). The mutual exclusion among monitor procedures and the immediate resumption of a signalled process ensure that the intended meaning holds since (i) no other process can access *fork* as long as a process is in the monitor, and thus (ii) the process that notes that two forks are available can update the status of $fork[i-1]$ and $fork[i+1]$ before its neighbors can continue.

Freedom from deadlock can be shown as follows. Let *eating = the number of philosophers eating*. Then when no process is in a monitor procedure, it is easy to see that

$$I=(\sum_{i:1}^{4} fork[i]=10-2*eating)$$

is invariant.

1. Initially *eating* = 0 and $\sum fork[i]=10$.
2. $\sum fork[i]$ is decreased by 2 and *eating* is incremented by 1 by every process completing *takefork*.
3. $\sum fork[i]$ is increased by 2 and *eating* is decremented by 1 by every process completing *releasefork*.

For deadlock to occur, we must have all processes waiting, and so *eating* = 0. By *I*, $\sum fork[i]=10$. Thus the last process, say *j*, to execute *takefork* and *wait* must have found $fork[j]=2$ and would not have waited, contrary to the assumption.

Consider now the scenario in Fig. 7.4.

Philosopher 2 is going to starve to death because philosophers 1 and 3 are conspiring to cause a lockout. By our rules, if there is even one scenario that leads to lockout, the solution must be rejected. Note that we do not assume that a philosopher eats indefinitely as eating has the status of a critical section. However, we can assume that two philosophers think very fast and eat very slowly.

Action	fork[0]	fork[1]	fork[2]	fork[3]	fork[4]
Initially	2	2	2	2	2
takefork[1]	1	2	1	2	2
takefork[3]	1	2	0	2	1
takefork[2]					
and wait	1	2	0	2	1
releasefork[1]	2	2	1	2	1
takefork[1]	1	2	0	2	1
releasefork[3]	1	2	1	2	2
takefork[3]	2	2	0	2	1

Fig. 7.4.

7.4 A CORRECT SOLUTION

```
program    diningphilosophers;
var        fork: array [0 . . 4] of (* binary *) semaphore;
           room: semaphore;
           i: integer;
procedure philosopher(i: integer);
begin
  repeat
    think;
    wait(room);
    wait( fork[i] );
    wait( fork[(i+1) mod 5] );
    eat;
    signal( fork[i] );
    signal( fork[(i+1) mod 5] );
    signal(room)
  forever
end;
begin (* main program *)
  room := 4;
  for i := 0 to 4 do fork[i] := 1;
  cobegin
    philosopher(0);
    philosopher(1);
    philosopher(2);
    philosopher(3);
    philosopher(4)
  coend
end.
```

Fig. 7.5.

The solution in Fig. 7.5 is similar to the first attempted solution except for the additional enclosing semaphore *room*. The safety properties hold as before. Deadlock is not a problem since *room* ensures that at most 4 philosophers are attempting to access forks. By a simple application of the "pigeon-hole principle" any attempt to distribute the five forks in the circle among the 4 philosophers must result in at least one philosopher having two forks. The semaphore invariant for the *room* is: *room* + (*number of processes between wait(room) and signal(room)*) = 4.

Let us now prove a series of lemmas that imply that this solution is starvation-free.

Lemma 7.1 If P_i executes *wait(fork[i])*, eventually it will complete the *wait*.

Proof
If P_i does not complete the *wait* it is only because *fork[i]* = 0 which implies that P_{i-1} is eating (since this is P_{i-1}'s right-hand fork which was taken just before *eat*). Eventually P_{i-1} finishes eating and executes *signal(fork[i])* allowing P_i to proceed.

(*Advanced*) *Remark* We must assume some definition of semaphores (such as Morris') stronger than the weak definition to avoid elementary semaphore lockout even among two processes. What we are interested in is showing that lockout cannot be caused by conspiring processes. The previous attempt suffered from lockout even though the monitor uses the strong FIFO assumption on its queues.

Lemma 7.2 If P_i is waiting indefinitely on *fork[i+1]* then P_{i+1} is waiting indefinitely on *fork[i+2]*.

Proof
Only P_i and P_{i+1} "compete" for the semaphore *fork[i+1]*. If P_{i+1} is terminated in *think* then *fork[i+1]* cannot block P_i. Similarly, P_i and P_{i+1} cannot be simultaneously blocked on the same semaphore *fork[i+1]* (think of the semaphore invariant). Thus if P_i is blocked on *fork[i+1]* and P_{i+1} is assumed never to *signal* that semaphore, the only possibility left is for P_{i+1} to be blocked indefinitely on the other semaphore: *fork[i+2]*.

Lemma 7.3 If P_i executes *wait(fork[i+1])*, eventually it will complete the *wait* (and eat).

Proof
By four successive applications of Lemma 7.2 we have that if P_i waits indefinitely on *fork[i+1]* then P_{i+j} waits indefinitely on *fork[i+j+1]*, $j = 1, \ldots, 4$ but this contradicts the semaphore invariant for *room*.

7.5 CONDITIONAL CRITICAL REGIONS

A solution to the dining philosophers problem using semaphores is difficult
to write because of the limited semantic content of the semaphore operations. A process can only test if an integer variable is zero and if so the
process is suspended. Thus we have no way of making a compound test on
the value of two "forks".

A synchronization primitive that does not suffer from this limitation is
the *conditional critical region*. A critical region is a primitive for mutual
exclusion. **region** *r* **begin** $s_1, \ldots, s_n$ **end** executes the sequence of statements
$s_1, \ldots, s_n$ as a critical section. If several processes try to enter the critical
region *r* simultaneously, then only one successfully enters and the others
must wait on a queue. When a process leaves a critical region, another
process from the queue is allowed to enter. Similarly, if a process attempts to
enter a region while another process is within the region, the new process
must join the queue.

Several regions can be declared to allow distinct critical sections to be
entered simultaneously, just as several semaphores or monitors may be
declared.

The solution to the mutual exclusion problem is immediate (Fig. 7.6).

```
program   mutualexclusion;
procedure p₁;
begin
  repeat
    rem1;
    region r begin crit1 end
  forever
end;
procedure p₂;
begin
  repeat
    rem2;
    region r begin crit2 end
  forever
end;
begin (* main program *)
  cobegin
    p₁; p₂
  coend
end.
```

Fig. 7.6.

To solve more difficult problems, the conditional critical region is used. Competition for entry to the critical region is allowed only if an entry condition is satisfied. These are similar to the Ada guards.

The syntax is: **await** b **region** r **begin** $s_1; \ldots ; s_n$ **end.** The semantics are: b is first evaluated. If b is true then the process may enter the competition to enter the critical region (and of course, succeed in entering if the region is free). If b is false, the process must enter the queue of waiting processes.

Whenever a process exits a critical region, all the processes waiting on its queue are released. Those processes having entry conditions must re-evaluate them. Processes with true conditions (including processes with no conditions) then compete to enter the critical region. The implementation should ensure that this competition is "fair", though FIFO is not specified as it is in the monitor.

We let each process evaluate its own condition—thus allowing expressions of arbitrary complexity, including those using local variables and procedures. The price we pay for this flexibility is the overhead of evaluating such expressions repeatedly as the process competes for entry into the critical section.

If we compare conditional critical regions with monitors, we see that the monitor has the flexibility of evaluating arbitrary expressions, but since the "condition" is named and explicitly signalled there is no busy wait overhead incurred by repeated evaluation.

Semaphores can be easily simulated:

$wait(sem)$: **await** $sem > 0$ **region** r **begin** $sem := sem - 1$ **end.**

$signal(sem)$: **region** r **begin** $sem := sem + 1$ **end.**

In Fig. 7.7 is a solution to the problem of the dining philosophers using conditional critical regions. The solution is similar to the monitor solution (Fig. 7.3). The conditional critical region primitive is here exactly what is needed to accomplish the compound test, something we could not do with semaphores.

```
program   diningphilosophers;
var       forks: array[0 . . 4] of integer;
          i: integer;
procedure philosopher(i: integer);
begin
  repeat
    think;
    await forks[i]=2
    region r begin
      forks[(i+1) mod 5] := forks[(i+1) mod 5]−1;
      forks[(i−1) mod 5] := forks[(i−1) mod 5]−1
    end;
```

```
      eat;
      region r begin
        forks[(i+1) mod 5] := forks[(i+1) mod 5]+1;
        forks[(i-1) mod 5] := forks[(i-1) mod 5]+1
      end
    forever
  end;
  begin (* main program *)
    for i := 0 to 4 do forks[i] := 2;
    cobegin
      philosopher(0);
      philosopher(1);
      philosopher(2);
      philosopher(3);
      philosopher(4)
    coend
  end.
```

Fig. 7.7.

Conditional critical regions do not have the theoretical appeal of the elementary semaphores. Practitioners have overwhelmingly preferred to implement the monitor. The effort involved is fully repaid by the flexibility of the concept and by the benefits obtained by the structuring of the data and procedures within a monitor. The Ada programming language, though it uses a different primitive, also encourages encapsulation of data and procedures in packages and tasks.

7.6 EXERCISES

7.1 Prove in greater detail the safety properties of the various algorithms.

7.2 Write one (or several) of the solutions in the chapter using the Ada rendezvous.

7.3 Program the following algorithm (in any formalism): a philosopher wishing to eat picks up his left fork; if his right fork is available, he picks it up and commences eating else he releases his left fork and repeats this cycle. Discuss the correctness of this algorithm.

7.4 All the solutions in this chapter are symmetrical, that is, all the philosophers execute the same code parameterized by the process number and furthermore no process explicitly uses its process number in the code. Try to find asymmetrical solutions: program them and discuss correctness. For example: change the first attempted solution so that one of the processes executes $wait(fork[i+1])$ and then $wait(fork[i])$ instead of conversely.

7.5 *Discuss the solution sketched in Fig. 7.8. We have deviated from Ada by allowing a **select** statement with no **accept** clause. The effect is that of a non-deterministic choice between the two possibilities.

```
task body fork is
begin
  loop
    select
      (* become i's left fork *)
      leftfork(i);
      accept releaseleftfork(i);
    or
      (* become i+1's right fork *)
      rightfork(i+1);
      accept releaserightfork(i+1);
    end select;
  end loop;
end;

task body philosopher is
begin
  loop
    think;
    select
      accept leftfork(i);
      accept rightfork(i);
    or
      accept rightfork(i);
      accept leftfork(i);
    end select;
    eat;
    releaseleftfork(i);
    releaserightfork(i);
  end loop;
end.
```

Fig. 7.8.

APPENDIX: IMPLEMENTATION KIT

A.1 INTRODUCTION

Soon, perhaps, every student of computer science will have his own minicomputer and sufficient time to learn its hardware thoroughly so that he may exercise concepts of concurrent programming by building a real-time or operating system from scratch. Until then, the accepted solution to class exercise of concurrent programming is to simulate concurrent execution.

There are several serious implementations which are noted in the references. This appendix contains the listing and documentation of a very simple system which can exercise concurrent execution with synchronization by semaphores. This system can thus be used by any instructor who does not have access to one of the more serious systems. The system is not efficient and is not intended to be used for extensive concurrent programming but it has been successfully used to demonstrate (the hard way) to students the difficulties of concurrent programming. Note that even though the system is written in Pascal, it uses only a subset of the language that could easily be translated into any block structured language. Similarly, the machine dependencies (such as record packing to save space) are clearly noted and easily changed. This appendix, however, presumes a knowledge of Pascal.

The program in the listing (Section A.8) is a simplification and modification of the Pascal-S interpreter originally written by N. Wirth. Pascal-S compiles a subset of Pascal into pseudo-code (P-code) for a hypothetical machine and then proceeds to interpret (simulate the execution of) this code. The general structure of Pascal-S is shown in Fig. A.1.

```
program pascals;
var code: array [1 . . codemax] of instructions;
procedure block;
```

119

begin
    Compile the Pascal-S program and store the compiled instructions in the array code.
end;
procedure *interpret*;
var *stack*: **array** [1 . . *stackmax*] **of** *integer*;
begin
    Simulate the instructions in code.
    The array stack serves as the memory of the simulated computer.
end;
begin (* *main program* *)
    initialize;
    block;
    interpret;
end.

Fig. A.1.

The object language P-code is for a stack machine. By this is meant that there are no registers or accumulators as on most computers. Instead, all operands are contained on a stack. A command such as Add needs no further specification since it automatically refers to the operands on the stack: Add the top two elements and replace them with the result as the top element in the stack. This architecture has actually been used on real computers (such as the Burroughs 6700) and even on pocket calculators (of the HP series).

In a stack machine $A := B + C$ is compiled into the following sequence of instructions:

1. Load the address of A onto the stack.
2. Load the value of B onto the stack.
3. Load the value of C onto the stack.
4. Add: remove the top two elements from the stack. Add them together and store the result as the top element on the stack.
5. Store the top element in the address that appears just below it on the stack. Remove these two elements.

A.2 THE COMPILER

The subset of Pascal compiled is:

1. Simple data types: *integer, boolean, char.*
2. Structures: **arrays** (including multidimensional arrays)
3. Strings: Only in the form: *write(# this is a title #).*
4. Operators: integer $(+, -, *, $ **div, mod**); boolean (**not, and, or**).

5. Relations: $=, <>, >, <, >=, >=$.

6. Declarations: **const, type, var, procedure, function**.

7. Statements:
   assignment;
   if b **then** s;
   if b **then** s_1 **else** s_2;
   while b **do** s;
   repeat s_1 . . . ; s_k **until** b;
   for $i := e_1$ **to** e_2 **do** s;

8. Block structure: procedures and functions may be nested under the usual block structuring rules of Pascal. They may have both variable and value parameters.

9. Concurrency: in the main program, a single compound statement of the form **cobegin** s_1; . . . ; s_n **coend** is allowed to indicate concurrent execution of s_1, . . . , s_n which must be global procedure calls.

10. *wait(s)* and *signal(s)* are predefined semaphore operations looking for integer variables. The initial value of the semaphore should be set by an assignment statement in the main program. For pedagogical reasons, a *semaphore* type is provided. However, it is synonymous with *integer* and provides no protection.

The compilation is by top-down recursive descent. A similar compiler is extensively described in the book *Structured System Programming* by Welsh and McKeag (1980) and thus we do not discuss it further except as needed to understand the concurrent interpreter. The only point we feel obliged to note is the use of a simpler data structure for the identifiers. They are kept in an array *tab*; the entry for an identifier contains a *link* field which contains the index of the previous identifier in the same level of block nesting.

A.3 THE P-CODE

The P-code instructions consist of an instruction field and possibly two operand fields x and y. x is used to point to the static level as described below and y is used to pass operands to the instructions. The interpreter uses the following data passed to it by the compiler.

btab—the block table, discussed below.

atab—the array table which has an entry for each array. The entries contain the following fields:
   low, high: the limits on the array index,
   elsize: the size of an array element.
   The following fields are used only by the compiler and will not concern us further:

size: the total size of the array,

eltyp, *inxtype*: the types of the elements and the index,

elref: a pointer to an entry in *atab* if the elements of the array are themselves arrays.

code—the compiled P-code.

stab—a table containing all the strings in the program. A string is identified by a starting index in *stab* and the number of characters.

The "memory" of the interpreter is the array *s* which is treated as a stack. All data are stored as integers. Characters may be stored and retrieved using Pascal functions *ord* and *chr*. For booleans we have written functions *booleantointeger* and *integertoboolean*.

The pseudo-instructions are summarized in the following table: *x* and *y* refer to the operand fields of the instruction and *t* is an index in *s* for the top of the stack. Instructions 0–7, 18–19, 32–33 will be further discussed later; the others are straightforward.

0: Load Address

1: Load Value

2: Load Indirect

3: Update Display

4: Cobegin

5: Coend

6: Wait

7: Signal

8: End of Line and End of File

10: Jump to address *y*

11: Jump to address *y* if $s[t]$ is false

14: Precedes the code of *s* in the **for**-loop:

for $i := e_1$ **to** e_2 **do** *s*.

y=the address of the instruction after the loop

$s[t] = e_2$; $s[t-1] = e_1$; $s[t-2] =$ address of *i*

15: Follows the code of *s* in the **for**-loop. y=address of *s*

18: Markstack

19: Procedure Call

21: Select an element of array *atab*[*y*] and push it onto the stack.

$s[t]$=value of the index

22: Push *y* words from address $s[t]$ onto the stack

23: Copy *y* words from address $s[t]$ to address $s[t-1]$

24: Push the literal *y* onto the stack

27: Read a value of type *y* into the variable whose address is $s[t]$

28: Write $s[t]$ characters from *stab*[*y*]

29: Write $s[t]$

31:   Halt
32:   Return from procedure
33:   Return from function
34:   $s[t]$ is an address; replace it by its value
38:   Store $s[t]$ at address $s[t-1]$
35–36, 45–59: Arithmetic and boolean operations
62, 63: Readln, Writeln
Others: omitted in the simplification from Wirth's Pascal-S

In a block-structured language the address of a variable is denoted by a pair
(Level, Address) where Level is the depth of nesting of blocks when the
variable was declared and Address is the offset of this variable within the
memory segment associated with this level. Note that the rules of block
structuring require that blocks be nested. Thus the memory segments can be
stacked. To access a non-local variable, one simply follows the links which
define the nesting until the proper level is reached (Fig. A.2, where s is a
local procedure of r which in turn is local to main program p). These links are
called *static links*.

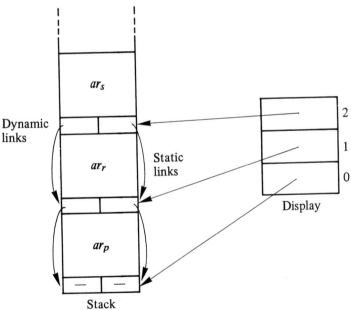

Fig. A.2.

If, however, we are going to execute the P-code, we need some faster
way of accessing a variable. This is done by using a *display* which is an array
indexed by levels whose elements point to the memory segments represent-
ing the current situation of block nesting. This is the same as saying that the

display holds the current static links. The address of a variable is calculated by computing $display[x]+y$ where $(x, y,)=$ (Level, Address).

Instructions 0, 1, 2, in the interpreter can now be understood as 0: Load an address; 1: Load a value; and 2: Indirect loading of a value. 2 is used in the case of a **var** parameter where what is passed to the procedure is the address of the address of a variable.

There is a price that must be paid for the elegance of the display. Any change in the current block structure, i.e. a procedure or function call must update the display. Hopefully, this occurs infrequently relative to accesses to the variables.

A.4   PROCEDURE CALLS

During the normal execution of a program by the interpreter, the two main pieces of dynamic information that must be maintained are *pc* (program counter) which points to the next instruction to be executed, and *t* which points to the current top of the stack. In addition, we maintain *b* which points to the bottom of the current stack segment and *stacksize* which points to the limit of the area allocated for the stack. Finally *display* keeps track of the addressing by nesting level.

Each time a procedure or function is called we need to allocate memory for the local variables as well as for certain additional information, such as the return address to jump to upon completion of the procedure. It is convenient to allocate this memory on the same stack that is used for the operands and to do the necessary adjustments upon the procedure call and return. The memory allocated upon procedure call is called an *activation record*.

In Pascal-S an activation record has a fixed part of five words:

ar[0]=Function result;
ar[1]=Return address;
ar[2]=Static link;
ar[3]=Dynamic link;
ar[4]=Table pointer.

When a function call is completed, the value of the function is left in place of the entire activation record. Conveniently, then, the effect of a function call is to push its result onto the stack just as if it were a normal operand. The return address contains the address of the instruction to be executed upon procedure exit. The table pointer contains the index into the identifier table *tab* for the procedure name. The *tab* field *adr* contains the index into the *code* table of the procedure code and *ref* contains the index into the block table to be described shortly.

If a procedure p has called a procedure q then the dynamic link of q points to the start of the activation record of p. It is used upon procedure exit to reset the various stack pointers. The static link points to the start of the activation record of the procedure which textually encloses the called procedure. This defines the block structure of the language and can be either used directly to access variables or reflected in the display.

To understand the difference between a static and a dynamic link look at Figs. A.2 and A.3 which show the activation records at two points during the execution of the following program.

program p;
procedure q;
begin ... **end**;
procedure r;
   procedure s;
   begin ... q ... **end**;
begin ... s ... **end**;
begin (* *main program* *) ... r ... **end**.

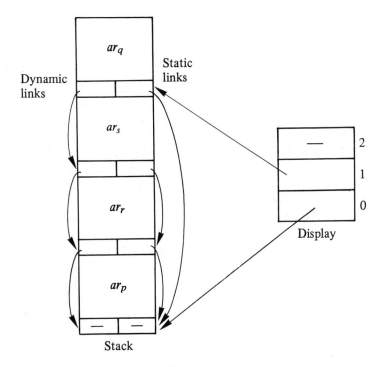

Fig. A.3

In Fig. A.2, p has called r (which is local to p) and then r has called s (which is local to r). The dynamic links show the sequence of the calls. The static links are the same since s can access the variables of r (which called it) and in turn r can access the variables of p (which called it). The display reflects the nesting levels 0, 1, 2.

In Fig. A.3, s has now called q, which is local to p and q cannot access the variables of r and s. This is indicated by the static pointer of the activation record of q which is now pointing to the activation record of p. Note that though the sequence of calls is longer, the display is now shorter.

As a check to see if you understand these concepts, draw the links for some recursive procedure *rec*. There will be a large number of different dynamic links—one for each recursive call of *rec*—but the static links will all point to the same activation record: that of the procedure in which *rec* is embedded.

We now describe the mechanics of a procedure call.
A procedure or function call is compiled into:

    18:   Markstack;
    Compile the parameters;
    19:   Call;
     3:   Update Display (if necessary).

The activation record consists of:

(1) the five word fixed area;
(2) an area for the actual parameters (whether values or addresses), and
(3) an area for the local variables of the procedure.

The array *btab* contains an entry for each procedure with fields: *psize* which is the sum of the lengths of areas (1) and (2) and *vsize* which is the sum of *psize* and the length of area (3). *vsize* is thus the total size of the activation record for this procedure. (There are also two fields *last* = pointer to the last identifier in this procedure and *lastpar* = pointer to the last parameter in this procedure. These are used only during the compilation and not during the simulation). *btab*[1] is an entry for the "environment" block containing predefined identifiers like *integer*. *btab*[2] is thus the entry for the main program.

The instruction Markstack is passed the index of the procedure name in the identifier table *tab*. Since the actual parameter evaluation may involve arbitrarily complicated computation, the main purpose of the Markstack instruction is to advance the stack pointer by five words so as to leave room on the stack for the fixed area. Then the evaluation of the parameters may freely use the stack above the fixed area.

In addition, Markstack performs the following services. Since the procedure name is seen and compiled before the actual parameters, the procedure name (via a pointer to the indentifier table) is conveniently passed at this point to the interpreter. Markstack stores this operand in $ar[4]$ for subsequent use by the Call. Markstack also checks that there is sufficient room in the stack for the activation record ($vsize$) and temporarily stores $vsize$ in the slot to be used by the dynamic link.

Now the actual parameters can be computed. The compiler has determined that when this has been completed, the stack pointer t will point to $psize$ words beyond the start of the activation record and in fact $psize$ (actually $(v/p)size - 1$ is stored to use as an offset) is passed as the operand to Call. The stack at the beginning of Call is shown in Fig. A.4.

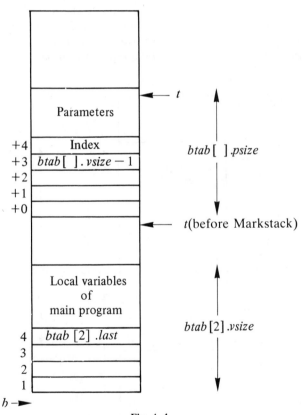

Fig. A.4.

With a slightly more descriptive notation than the bare listing, the code of Call is shown in Fig. A.5. Fig. A.6. shows the stack following the completion of the procedure call. Procedure exit is relatively simple (Fig. A.7).

$oldb := b;$ (* Save the previous bottom of stack for the dynamic link *)

$b := t - y;$ (* New bottom stack is computed from *psize* as passed in y *)

$inx := s[b+4];$ (* Retrieve index of procedure in *tab* as left by Markstack *)

$t := s[b+3]+b;$ (* Compute new top of stack from *vsize* left by Markstack *)

$level := tab[inx].lev;$ (* Get nesting level in which procedure is defined *)

$s[b + 2] := display[level];$ (* Static link is the bottom of the stack segment of the level of definition *)

$display[level+1] := b;$ (* The procedure executes one level up, so update the *display* *)

$s[b + 3] := oldb;$ (* Set the dynamic link *)
$s[b + 1] := pc;$ (* Return address *)
$pc := tab[inx].adr;$ (* Jump to the procedure code *)

Fig. A.5.

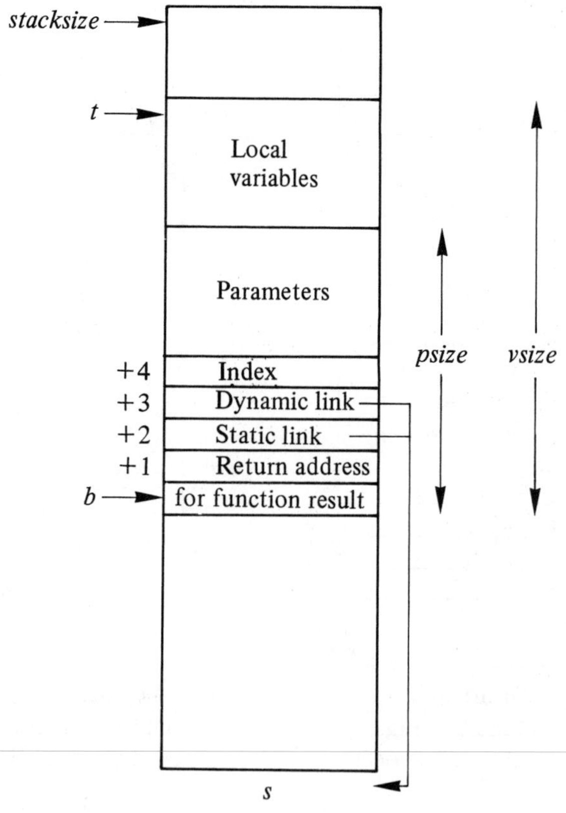

Fig. A.6.

$t := b - 1;$ (* Restore the old top of the stack. For
                                        a function exit this is $t := b$; *)

$pc := s[b + 1];$ (* Restore return address *)
$b := s[b + 3];$ (* Restore bottom of stack from the
                                        dynamic link *)

Fig. A.7.

There is one more problem that must be solved. If the procedure call is to a procedure on a lower static level than that of the calling procedure, the display will not be correct upon exit. In Fig. A.3 when the execution of q terminates, only $display[0]$ pointing to the activation record of the main program will be correct. $display[1]$ and $display[2]$ must be made to point to the activation records of r and s, respectively. This is done by the interpreter instruction (3): Update display(x, y) which updates $display[y]$ down to $display[x]$ by following the static links. In this case, the compiler must insert Update $display(1, 2)$ which will cause b—currently pointing to the bottom of the activation record of s to be stored in $display[2]$ and then obtain from the static link the index of the bottom of the activation record of r to store in $display[1]$.

This completes our discussion of Pascal-S as applied to sequential programs. Before reading further you might want to think of the problems that will be encountered in extending the system to concurrent execution.

A.5 CONCURRENCY

To execute several processes concurrently we need to maintain: a stack for each process which will be used for the local data of the process (including the data of any procedures called by the process) and a set of register images for each process. To execute a particular process, its images are loaded into the physical registers before execution.

In the case of Pascal-S "registers" are pc, the stack pointers b, t and $stacksize$ and $display$. We maintain a table $ptab$† with one entry for each process containing this information as well as two further pieces of information: $active$ indicating whether this process is active (as opposed to terminated) and a pointer $suspend$ which points to a semaphore if the process is suspended on a semaphore. The register images are "loaded" into the interpreter by the **with** $ptab[curpr]$ statement which makes all references to the above variables take their values from the entry for the current process—$curpr$.

† The constant $pmax$ controls the size of $ptab$ and hence the number of concurrent processes allowed. For most efficient use of central memory, it is best to tailor a version of the kit for each class exercise. Conway's problem (Exercise 4.2) runs with $pmax = 3$ but Parnas' problem (Exercise 4.15) needs $pmax = 9$.

Similarly we should have a set of stacks—one for the main process (which contains all the global variables) and one for each concurrent process. If this is done, however, the interpreter would have to be passed an index to the stack table on each memory access. A simpler solution is to divide the single physical stack (the array *s*) into several logical segments—one for each process. Then as long as the displays are correct, all accessing is done normally. See Fig. A.8.

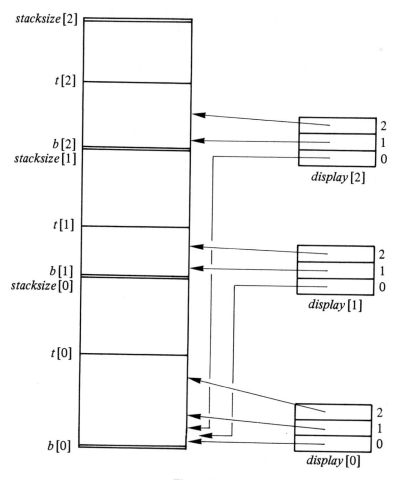

Fig. A.8.

For simplicity, the stacks are pre-allocated. The use of dynamic allocation does not involve any conceptual difficulties but for our purposes it seems unnecessary. Various versions of this system can be tailored to suit each assigned exercise.

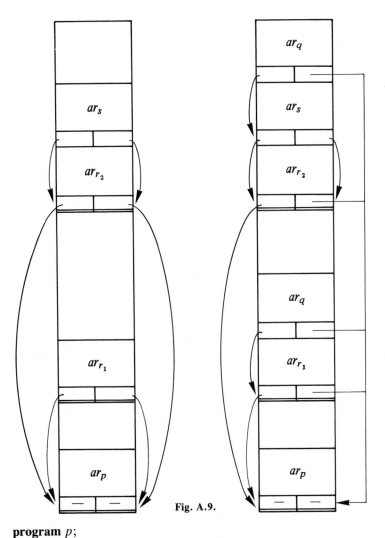

Fig. A.9.

Fig. A.10.

```
program p;
procedure q;
begin . . . end;
procedure r₁;
begin . . . q . . . end;
procedure r₂;
   procedure s;
   begin . . . q . . . end;
begin . . . s . . . end;
begin (* p *)
   cobegin r₁; r₂ coend
end.
```

The semantics of **cobegin** . . . **coend** are as follows: The concurrent processes are prepared for execution: Their parameters are evaluated and stored in the stack segments and the values of their register images are computed. Upon execution of the **coend** statement, the main process is set to inactive and the concurrent processes are activated. When all the concurrent processes have terminated (become inactive) the main process is reactivated.

This is accomplished by preferring the execution of the main process if it is active (**if** $ptab[0].active$ **then** $curpr := 0$ **else** . . .) and by having **coend** set the main process to inactive. At this point the choice of the current process to execute will always be made among the concurrent processes. The exit procedure instruction performed for a concurrent process will decrement the number of active processes and if this count reaches zero the main process is set to active.

The implementation of **cobegin** is to have it set a flag ($pflag$) which is used during the calling sequence to decide whether this is a concurrent process call. If so, the Markstack instruction allocates a $ptab$ entry (by incrementing npr) and marks the stack segment assigned for the new process. This stack segment was defined during initialization of $ptab$. During the actual parameter evaluation, $pflag$ ensures that the stack segment upon which the parameters are pushed is in fact this new segment (**if** $pflag$ **then** . . .). In the Call, the only difference is that the initialized activation record happens to belong to the called process and not the main process. Upon execution of the **coend**, $pflag$ is reset so that if a concurrent process calls a procedure, the normal calling sequence is done.

Exit procedure uses a dummy return address to detect that this is the termination of a concurrent process. It deactivates the process and if this is the last concurrent process, it reactivates the main process.

Figure A.9 shows the stack at two instances during the execution of the program in Fig. A.10.

A.6   SEMAPHORES

The interpreter allows easy implementation of semaphores because the entire operation can be done as one pseudo-instruction and does not require that the testing and incrementing be done as separate operations. $wait(s)$ and $signal(s)$ are compiled as standard procedures being passed the address of the parameter. $wait$ is implemented as: **if** $s > 0$ **then** $s := s - 1$ **else** $suspend :=$ address of s. $signal$ is implemented as : Search for a $ptab$ entry with $suspend$ = address of s. **if** found **then** $suspend := 0$ **else** $s := s + 1$.

Remark   This is the definition of semaphores suggested by Morris and is the one that has the most natural implementation.

A.7 RANDOMIZATION

A random number generator is used to give the illusion of concurrency. *chooseproc* searches from a random starting point in *ptab* for a process which is active and not suspended. If there are no such processes, you can let the interpreter run to the computer's time limit or, as shown, explicitly declare a deadlock to the student.

Randomization is also used in *signal* to ensure that the awakened process will not be predetermined. *stepcount* is used to reduce the number of executions of *chooseproc* by allowing a process to execute a (random) number of consecutive steps before a process switch is done. Note that if a process becomes inactive or suspended, then *stepcount* is zeroed to force a process switch.

A.8 PROGRAM LISTING

```
program pascals(input, output);
(*author: N. Wirth, E. T. H. ch—8092 Zurich, 1. 3. 76*)
(*modified: M. Ben-Ari, Tel Aviv Univ., 1980*)
label 99;
const nkw=26;                          (*no. of key words*)
      alng=10;              (*no. of significant chars in identifiers*)
      llng=121;                         (*inputline length*)
      kmax=15;              (*max no. of significant digits*)
      tmax=70;                          (*size of table*)
      bmax=20;                          (*size of block-table*)
      amax=10;                          (*size of array table*)
      cmax=500;                         (*size of code*)
      lmax=7;                           (*maximum level*)
      smax=150;                         (*size of string-table*)
      omax=63;                          (*highest order code*)
      xmax=131071;                      (*2**17 – 1*)
      nmax=281474976710655;             (*2**48–1*)
      lineleng=132;                     (*output line length*)
      linelimit=400;                    (*max lines to print*)
      stmax=1400;                       (*stacksize*)
      stkincr=200;                      (*stacksize for each process*)
      pmax=3;                           (*max concurrent processes*)
```

```
type symbol=(intcon, charcon, string,
             notsy, plus, minus, times, idiv, imod, andsy, orsy,
             eql, neq, gtr, geq, lss, leq,
             lparent, rparent, lbrack, rbrack, comma, semicolon,
             period,
             colon, becomes, conststy, typesy, varsy, functionsy,
             proceduresy, arraysy, programsy, ident,
             beginsy, ifsy, repeatsy, whilesy, forsy,
             endsy, elsesy, untilsy, ofsy, dosy, tosy, thensy);
```

$index = -xmax \, .. \, +xmax;$
$alfa = $ **packed array** $[1 \, .. \, alng]$ **of** $char;$
$object = (konstant, variable, type1, prozedure, funktion);$
$types = (notyp, ints, bools, chars, arrays);$
$er = (erid, ertyp, erkey, erpun, erpar, ernf, erdup, erch, ersh, erln);$
$symset = $ **set of** $symbol;$
$typset = $ **set of** $types;$
$item = $ **record**

   $typ: types; ref: index;$
  end;
$order = $ **packed record**

   $f: -omax \, .. \, +omax;$
   $x: -lmax \, .. \, +lmax;$
   $y: -nmax \, .. \, +nmax;$
  end;

var $sy: symbol;$ *(*last symbol read by insymbol*)*
 $id: alfa;$ *(*identifier from insymbol*)*
 $inum: integer;$ *(*integer from insymbol*)*
 $rnum: real;$ *(*real number from insymbol*)*
 $sleng: integer;$ *(*string length*)*
 $ch: char;$ *(*last character read from source program*)*
 $line: $ **array** $[1 \, .. \, llng]$ **of** $char;$
 $cc: integer;$ *(*character counter*)*
 $lc: integer;$ *(*program location counter*)*
 $ll: integer;$ *(*length of current line*)*
 $errs: $ **set of** $er;$
 $errpos: integer;$
 $progname: alfa;$
 $skipflag: boolean;$
 $constbegsys, typebegsys, blockbegsys, facbegsys, statbegsys:$
  $symset;$
 $key: $ **array** $[1 \, .. \, nkw]$ **of** $alfa;$
 $ksy: $ **array** $[1 \, .. \, nkw]$ **of** $symbol;$
 $sps: $ **array** $[char]$ **of** $symbol;$ *(*special symbols*)*

```
    t, a, b, sx, c₁, c₂: integer;                    (* indices to tables*)
    stantyps: typset;
    display: array [0 . . lmax] of integer;

    tab: array [0 . . tmax] of                       (*identifier table*)
            packed record
                name: alfa; link: index;
                obj: object; typ: types;
                ref: index; normal: boolean;
                lev: 0 . . lmax; adr: integer;
            end;
    atab: array [1 . . amax] of                      (*array-table*)
            packed record
                inxtyp, eltyp: types;
                elref, low, high, elsize, size: index;
            end;
    btab: array [1 . . bmax] of                      (*block-table*)
            packed record
                last, lastpar, psize, vsize: index
            end;
    stab: packed array [0 . . smax] of char;         (*string table*)
    code: array [0 . . cmax] of order;

procedure errormsg;
    var k: er;
        msg: array [er] of alfa;
begin
    msg[erid] := #identifier#; msg[ertyp] := #type      #;
    msg[erkey] := #keyword #; msg[erpun] := #punctuatio#;
    msg[erpar] := #parameter #; msg[ernf] := #not found #;
    msg[erdup] := #duplicate#; msg[erch] := #character #;
    msg[ersh] := #too short #; msg[erln] := #too long  #;
    message(# compilation errors#);
    writeln; writeln(# key words#);
    for k := erid to erln do if k in errs then
        writeln(ord(k), # #,msg[k])
end (*errormsg*);

procedure endskip;
begin (*underline skipped part of input*)
    while errpos < cc do
        begin write(#-#); errpos := errpos+1
        end;
    skipflag := false
end (*endskip*);
```

```
procedure nextch; (*read next character; process line end*)
begin if cc=ll then
      begin if eof(input) then
         begin writeln;
            writeln(# program incomplete#);
            errormsg; goto 99
         end;
      if errpos <> 0 then
         begin if skipflag then endskip;
            writeln; errpos := 0
         end;
      write(lc:5, # #);
      ll := 0; cc ·= 0;
      while not eoln(input) do
         begin ll := ll+1; read(ch); write(ch); line[ll] := ch
         end;
      writeln; ll := ll+1; read(line[ll])
      end;
   cc := cc+1; ch := line[cc];
end (*nextch*);

procedure error(n: er);
begin if errpos = 0 then write(# ****#);
   if cc > errpos then
      begin write(# #: cc–errpos, #'#, ord(n):2);
         errpos := cc+3; errs := errs+[n]
      end
end (*error*);

procedure fatal(n: integer);
   var msg: array [1 .. 6] of alfa;
begin writeln; errormsg;
   msg[ 1] := #identifier#; msg[ 2] := #procedures#;
   msg[ 3] := #strings  #; msg[ 4] := #arrays   #;
   msg[ 5] := #levels   #; msg[ 6] := #code     #;
   writeln(# compiler table for #, msg[n], # is too small#);
   goto 99 (* terminate compilation*)
end (*fatal*);

(* ------------------------------------------------------------------------insymbol-*)

procedure insymbol; (*reads next symbol*)
   label 1,2,3;
   var i, j, k, e: integer;
```

```
begin (*insymbol*)
1: while ch=# # do nextch;
   case ch of
#a#, #b#, #c#, #d#, #e#, #f#, #g#, #h#, #i#,
#j#, #k#, #l#, #m#, #n#, #o#, #p#, #q#, #r#,
#s#, #t#, #u#, #v#, #w#, #x#, #y#, #z#:
   begin (*identifier or wordsymbol*) k := 0; id := #          #;
      repeat if k < alng then
               begin k := k+1; id[k] := ch
               end;
            nextch
      until not (ch in [#a#. . #z#, #0#. . #9#]);
      i := 1; j := nkw; (*binary search*)
      repeat k := (i+j) div 2;
         if id <= key[k] then j := k-1;
         if id >= key[k] then i := k+1
      until i > j;
      if i-1 > j then sy := ksy[k] else sy := ident
   end;
#0#, #1#, #2#, #3#, #4#, #5#, #6#, #7#, #8#, #9#:
   begin (*number*) k := 0; inum := 0; sy := intcon;
      repeat inum := inum*10 + ord(ch) - ord(#0#);
         k := k+1; nextch
      until not (ch in [#0#. . #9#]);
      if (k > kmax) or (inum > nmax) then
         begin error(erln); inum := 0; k := 0
         end;
   end;

#:#, col: begin nextch;
            if ch = #=# then
               begin sy := becomes; nextch
               end else sy := colon
         end;
#<# : begin nextch;
         if ch = #=# then begin sy := leq; nextch end else
         if ch = #># then begin sy := neq; nextch end else
            sy := lss
      end;
#># : begin nextch;
         if ch = #=# then begin sy := geq; nextch end else
            sy := gtr
      end;
#.# : begin nextch;
         if ch = #.# then
```

```
                 begin sy := colon; nextch
                 end else sy := period
              end;
    ####: begin k := 0;
        2:  nextch;
             if ch = #### then
                begin nextch; if ch <> #### then goto 3
                end;
             if sx+k = smax then fatal(3);
             stab[sx+k] := ch; k := k+1;
             if cc = 1 then
                begin (*end of line*) k := 0;
                end
             else goto 2;
        3:  if k = 1 then
                begin sy := charcon; inum := ord(stab[sx])
                end else
             if k = 0 then
             begin error(ersh); sy := charcon; inum := 0
             end else
                begin sy := string; inum := sx; sleng := k; sx := sx+k
                end
           end;
    #(#:  begin nextch;
             if ch <> #*# then sy := lparent else
             begin (*comment*) nextch;
               repeat
                  while ch <> #*# do nextch;
                  nextch
               until ch = #)#;
               nextch; goto 1
             end
           end;
    #+#, #-#, #*#, #)#, #=#, #,#, #[#, #]#, #;# :
       begin sy := sps[ch]; nextch
       end;
    #$#, #!#, #@#, #\#, #:#, #_#, #?#, #'#, #"#, #&#, #/# :
         begin error(erch); nextch; goto 1
         end
       end
    end (*insymbol*);
```

(* -- enter ---*)

```
procedure enter (x₀: alfa; x₁: object;
                    x₂: types; x₃: integer);
begin t := t+1; (*enter standard identifier*)
  with tab[t] do
  begin name := x₀; link := t-1; obj := x₁;
    typ := x₂; ref := 0; normal := true;
    lev := 0; adr := x₃
  end
end (*enter*);

procedure enterarray(tp: types; l,h: integer);
begin if l > h then error(ertyp);
  if (abs(l)>xmax) or (abs(h)>xmax) then
    begin error(ertyp); l := 0; h := 0;
    end;
  if a = amax then fatal(4) else
    begin a := a+1;
      with atab[a] do
        begin inxtyp := tp; low := l; high := h
        end
    end
end (*enterarray*);

procedure enterblock;
begin if b = bmax then fatal(2) else
  begin b := b+1; btab[b]. last := 0; btab[b]. lastpar := 0
  end
end (*enterblock*);

procedure emit(fct: integer);
begin if lc = cmax then fatal(6);
  code[lc]. f := fct; lc := lc+1
end (*emit*);
procedure emit1(fct, b: integer);
begin if lc=cmax then fatal(6);
  with code[lc] do
    begin f := fct; y := b end;
  lc := lc+1
end (*emit1*);
procedure emit2(fct, a, b: integer);
begin if lc = cmax then fatal(6);
  with code[lc] do
    begin f := fct; x := a; y := b end;
  lc := lc+1
end (*emit2*);
```

(* -- *block* --*)

procedure *block*(*fsys*: *symset*; *isfun*: *boolean*; *level*: *integer*);

  type *conrec* =
    record *tp*: *types*; *i*: *integer* **end**;

  var *dx*: *integer*;   (**data allocation index**)
    prt: *integer*;   (**t–index of this procedure**)
    prb: *integer*;   (**b–index of this procedure**)
    x: *integer*;

  procedure *skip*(*fsys*: *symset*; *n*: *er*);
  begin *error*(*n*); *skipflag* := *true*;
    while not (*sy* **in** *fsys*) **do** *insymbol*;
    if *skipflag* **then** *endskip*
  end (**skip**);

  procedure *test*(s_1, s_2: *symset*; *n*: *er*);
  begin if not (*sy* **in** s_1) **then**
    skip(s_1+s_2, *n*)
  end (**test**);

  procedure *testsemicolon*;
  begin
    if *sy* = *semicolon* **then** *insymbol* **else** *error*(*erpun*);
    test([*ident*]+*blockbegsys*, *fsys*, *erkey*);
  end (**testsemicolon**);

  procedure *enter*(*id*: *alfa*; *k*: *object*);
    var *j*, *l*: *integer*;
  begin if *t* = *tmax* **then** *fatal*(1) **else**
    begin *tab*[0]. *name* := *id*;
      j := *btab*[*display*[*level*]]. *last*; *l* := *j*;
      while *tab*[*j*]. *name* <> *id* **do** *j* := *tab*[*j*]. *link*;
      if *j* <> 0 **then** *error*(*erdup*) **else**
      begin *t* := *t*+1;
        with *tab*[*t*] **do**
        begin *name* := *id*; *link* := *l*;
          obj := *k*; *typ* := *notyp*; *ref* := 0; *lev* := *level*; *adr* := 0
        end;
        btab[*display*[*level*]]. *last* := *t*
      end
    end
  end (**enter**);

```pascal
function loc(id: alfa): integer;
  var i, j: integer; (*locate id in table*)
begin i := level; tab[0]. name := id; (*sentinel*)
  repeat j := btab[display[i]]. last;
    while tab[j]. name <> id do j := tab[j]. link;
    i := i-1;
  until (i<0) or (j<>0);
  if j = 0 then error(ernf); loc := j
end (*loc*);


procedure entervariable;
begin if sy = ident then
    begin enter(id, variable); insymbol
    end
  else error(erid)
end (*entervariable*);


procedure constant(fsys: symset; var c: conrec);
  var x, sign: integer;
begin c. tp := notyp; c.i := 0;
  test(constbegsys, fsys, erkey);
  if sy in constbegsys then
  begin
    if sy = charcon then
      begin c. tp := chars; c. i := inum; insymbol
      end
    else
      begin sign := 1;
        if sy in [plus, minus] then
          begin if sy = minus then sign := -1;
            insymbol
          end;
        if sy = ident then
          begin x := loc(id);
            if x <> 0 then
              if tab[x]. obj <> konstant then error(ertyp) else
              begin c. tp := tab[x]. typ;
                    c. i := sign*tab[x]. adr
              end;
            insymbol
          end
        else
        if sy = intcon then
```

```
              begin c. tp := ints ; c. i := sign*inum; insymbol end
                    else skip(fsys, erkey)
          end;
        test(fsys, [], erkey);
      end
end (*constant*);

  procedure typ(fsys: symset; var tp: types; var rf, sz: integer);
    var x: integer;
        eltp: types; elrf: integer;
        elsz, offset, t0, t1: integer;

    procedure arraytyp(var aref,arsz: integer);
      var eltp: types;
          low, high: conrec;
          elrf, elsz: integer;
    begin constant([colon, rbrack, ofsy]+fsys, low);
        if sy = colon then insymbol else error(erpun);
        constant([rbrack, comma, ofsy]+fsys, high);
        if high. tp <> low. tp then
          begin error(ertyp); high. i := low. i
          end;
        enterarray(low. tp, low. i, high. i); aref := a;
        if sy = comma then
          begin insymbol; eltp := arrays; arraytyp(elrf,elsz)
          end else
        begin
          if sy = rbrack then insymbol else error(erpun);
          if sy = ofsy then insymbol else error(erkey);
          typ(fsys, eltp, elrf, elsz)
        end;
        with atab[aref] do
        begin arsz := (high–low+1)*elsz; size := arsz;
          eltyp := eltp; elref := elrf; elsize := elsz
        end;
      end (*arraytyp*);

  begin (*typ*) tp := notyp; rf := 0; sz := 0;
    test(typebegsys, fsys, erid);
    if sy in typebegsys then
      begin
        if sy = ident then
```

```
      begin x := loc(id);
        if x <> 0 then
        with tab[x] do
          if obj <> type1 then error(ertyp) else
          begin tp := typ; rf := ref; sz := adr;
            if tp = notyp then error(ertyp)
          end;
          insymbol
        end else
        if sy = arraysy then
        begin insymbol;
          if sy = lbrack then insymbol else error(erpun);
          tp := arrays; arraytyp(rf, sz)
        end else
        test(fsys, [ ], erkey);
      end
end (*typ*);


procedure parameterlist; (*formal parameter list*)
    var tp: types;
        rf, sz, x, t0: integer;
        valpar: boolean;
begin insymbol; tp := notyp; rf := 0; sz := 0;
  test([ident, varsy], fsys+[rparent], erpar);
  while sy in [ident, varsy] do
    begin if sy <> varsy then valpar := true else
            begin insymbol; valpar := false
            end;
        t0 := t; entervariable;
        while sy = comma do
          begin insymbol; entervariable;
          end;
        if sy = colon then
          begin insymbol;
            if sy <> ident then error(erid) else
            begin x := loc(id); insymbol;
              if x <> 0 then
              with tab[x] do
                if obj <> type1 then error(ertyp) else
                  begin tp := typ; rf := ref;
                    if valpar then sz := adr else sz := 1
                  end;
```

```
            end;
            test([semicolon, rparent], [comma, ident]+fsys, erpun)
          end
      else error(erpun);
      while t₀ < t do
      begin t₀ := t₀+1;
        with tab[t₀] do
        begin typ := tp; ref := rf;
          normal := valpar; adr := dx; lev := level;
          dx := dx + sz
        end
      end;
      if sy <> rparent then
      begin if sy = semicolon then insymbol else error(erpun);
        test([ident, varsy], [rparent]+fsys, erkey);
      end
    end (*while*);
  if sy = rparent then
    begin insymbol;
      test([semicolon, colon], fsys, erkey);
    end
  else error(erpun)
end (*parameterlist*);

procedure constantdeclaration;
  var c: conrec;
begin insymbol;
  test([ident], blockbegsys, erid);
  while sy=ident do
    begin enter(id, konstant); insymbol;
      if sy=eql then insymbol else error(erpun);
      constant([semicolon, comma, ident]+fsys, c);
      tab[t]. typ := c. tp; tab[t]. ref := 0;
      tab[t]. adr := c. i;
      testsemicolon
    end
end (*constantdeclaration*);

procedure typedeclaration;
  var tp: types; rf, sz, t₁: integer;
begin insymbol;
  test([ident], blockbegsys, erid);
  while sy=ident do
```

```
          begin enter(id, type1); t₁ := t; insymbol;
            if sy=eql then insymbol else error(erpun);
            typ([semicolon, comma, ident]+fsys, tp, rf, sz);
            with tab[t₁] do
              begin typ := tp; ref := rf; adr := sz
              end;
            testsemicolon
          end
    end (*typedeclaration*);

    procedure variabledeclaration;
      var t₀, t₁, rf, sz: integer;
        tp: types;
    begin insymbol;
      while sy = ident do
      begin t₀ := t; entervariable;
        while sy = comma do
          begin insymbol; entervariable;
          end;
        if sy = colon then insymbol else error(erpun);
        t₁ := t;
        typ([semicolon, comma, ident]+fsys, tp, rf, sz);
        while t₀ < t₁ do
        begin t₀ := t₀+1;
          with tab[t₀] do
          begin typ := tp; ref := rf;
            lev := level; adr := dx; normal := true;
            dx := dx+sz
          end
        end;
        testsemicolon
      end
    end (*variabledeclaration*);

    procedure procdeclaration;
      var isfun: boolean;
    begin isfun := sy = functionsy; insymbol;
      if sy <> ident then
        begin error(erid); id := #        #
        end;
      if isfun then enter(id, funktion) else enter(id, prozedure);
      tab[t]. normal := true;
      insymbol; block([semicolon]+fsys, isfun, level+1);
```

```
        if sy=semicolon then insymbol else error(erpun);
        emit(32+ord(isfun)) (*exit*)
    end (*proceduredeclaration*);
```

(*-- statement--*)

```
procedure statement(fsys: symset);
  var i: integer; x: item;
  procedure expression(fsys: symset; var x: item); forward;

  procedure selector(fsys: symset; var v: item);
    var x: item; a, j: integer;
  begin
        if sy <> lbrack then error(ertyp);
        repeat insymbol;
          expression(fsys+[comma, rbrack], x);
          if v. typ <> arrays then error(ertyp) else
            begin a := v. ref;
              if atab[a]. inxtyp <> x. typ then error(ertyp) else
                emit1(21, a);
              v. typ := atab[a]. eltyp; v. ref := atab[a]. elref
            end
        until sy <> comma;
        if sy = rbrack then insymbol else error(erpun);
      test(fsys, [ ], erkey);
  end (*selector*);

  procedure call(fsys: symset; i: integer);
    var x: item;
        lastp, cp, k: integer;
  begin emit1(18, i); (*markstack*)
    lastp := btab[tab[i]. ref]. lastpar; cp := i;
    if sy = lparent then
    begin (*actual parameter list*)
      repeat insymbol;
        if cp >= lastp then error(erpar) else
        begin cp := cp+1;
          if tab[cp]. normal then
          begin (*value parameter*)
            expression(fsys+[comma, colon, rparent], x);
            if x. typ=tab[cp]. typ then
              begin
                if x. ref <> tab[cp]. ref then error(ertyp) else
                if x. typ = arrays then emit1(22,atab[x. ref]. size)
              end else if x. typ<>notyp then error(ertyp);
```

```
                  end else
begin (*variable parameter*)
                if sy <> ident then error(erid) else
                begin k := loc(id); insymbol;
                  if k <> 0 then
                  begin if tab[k]. obj <> variable then error(erpar);
                     x. typ := tab[k]. typ; x. ref := tab[k]. ref;
                     if tab[k]. normal then emit2(0, tab[k]. lev, tab[k].
                     adr)
                         else emit2(1, tab[k]. lev, tab[k]. adr);
                     if sy=lbrack then
                         selector(fsys+[comma, colon, rparent], x);
                        if (x. typ<>tab[cp]. typ) or (x. ref<>tab[cp]. ref)
                        then error(ertyp)
                  end
                end
              end
            end;
          test([comma, rparent], fsys, erkey);
        until sy <> comma;
        if sy = rparent then insymbol else error(erpun)
      end;
      if cp < lastp then error(erpar); (*too few actual parameters*)
      emit1(19, btab[tab[i]. ref]. psize–1);
      if tab[i]. lev < level then emit2(3, tab[i]. lev, level)
end (*call*);


function resulttype(a, b: types): types;
begin
  if (a>ints) or (b>ints) then
    begin error(ertyp); resulttype := notyp
    end else
  if (a=notyp) or (b=notyp) then resulttype := notyp else
    resulttype := ints
end (*resulttype*);


procedure expression;
  var y: item; op: symbol;

  procedure simpleexpression(fsys: symset; var x: item);
    var y: item; op: symbol;

    procedure term(fsys: symset; var x: item);
      var y: item; op: symbol; ts: typset;
```

```
     procedure factor(fsys: symset; var x: item);
     var i, f: integer;

begin (*factor*) x. typ := notyp; x. ref := 0;
  test(facbegsys, fsys, erpun);
  while sy in facbegsys do
    begin
      if sy = ident then
      begin i := loc(id); insymbol;
        with tab[i] do
        case obj of
konstant: begin x. typ := typ; x. ref := 0;
                emit1(24, adr)
            end;
variable: begin x. typ := typ; x. ref := ref;
              if sy = lbrack then
                begin if normal then f := 0 else f := 1;
                  emit2(f, lev, adr);
                  selector(fsys, x);
                  if x. typ in stantyps then emit(34)
                  end else
                  begin
                    if x. typ in stantyps then
                      if normal then f := 1 else f := 2
                    else
                      if normal then f := 0 else f := 1;
                    emit2(f, lev, adr)
                  end
              end;
type1, prozedure: error(ertyp);
funktion : begin x. typ := typ;
                if lev <> 0 then call(fsys, i)
                              else emit1(8, adr)
            end
          end (*case, with*)
        end else
      if sy in [charcon, intcon] then
        begin if sy = charcon then x. typ := chars
                              else x. typ := ints;
              emit1 (24, inum);
              x. ref := 0; insymbol
        end else
      if sy = lparent then
```

```
            begin insymbol; expression(fsys+[rparent], x);
               if sy = rparent then insymbol else error(erpun)
            end else
         if sy = notsy then
            begin insymbol; factor(fsys, x);
               if x. typ=bools then emit(35) else
                  if x. typ<>notyp then error(ertyp)
            end;
         test(fsys, facbegsys, erkey);
         end (*while*)
      end (*factor*);

begin (*term*)
   factor(fsys+[times, idiv, imod, andsy], x);
   while sy in [times, idiv, imod, andsy] do
      begin op := sy; insymbol;
      factor(fsys+[times, idiv, imod, andsy], y);
      if op = times then
      begin x. typ := resulttype(x. typ, y. typ);
         if x. typ = ints then emit(57)
      end else
      if op = andsy then
      begin if (x. typ=bools) and (y. typ=bools) then
               emit(56) else
            begin if (x. typ<>notyp) and (y. typ<>notyp)
                     then error(ertyp);
               x. typ := notyp
            end
      end else
      begin (*op in [idiv, imod]*)
         if (x. typ=ints) and (y. typ=ints) then
            if op=idiv then emit(58)
                       else emit(59) else
            begin if (x. typ<>notyp) and (y. typ<>notyp) then
                     error(ertyp);
               x. typ := notyp
            end
         end
      end
   end (*term*);

begin (*simpleexpression*)
   if sy in [plus, minus] then
```

```
            begin op := sy; insymbol;
              term(fsys+[plus, minus], x);
              if x. typ > ints then error(ertyp) else
                if op = minus then emit(36)
            end else
         term(fsys[plus, minus, orsy], x);
         while sy in [plus, minus, orsy] do
            begin op := sy; insymbol;
              term(fsys+[plus, minus, orsy], y);
              if op = orsy then
              begin
                if (x. typ=bools) and (y. typ=bools) then emit(51) else
                  begin if (x. typ<>notyp) and (y. typ<>notyp) then
                      error(ertyp);
                    x. typ := notyp
                  end
              end else
              begin x. typ := resulttype(x. typ, y. typ);
                if x. typ = ints then if op = plus then emit(52)
                                                   else emit(53)
              end
            end
         end (*simpleexpression*);


      begin (*expression*);
         simpleexpression(fsys+[eql, neq, lss, leq, gtr, geq], x);
         if sy in [eql, neq, lss, leq, gtr, geq] then
            begin op := sy; insymbol; simpleexpression(fsys, y);
              if (x. typ in [notyp, ints, bools, chars])
                and (x. typ = y. typ) then
                case op of
                  eql: emit(45);
                  neq: emit(46);
                  lss: emit(47);
                  leq: emit(48);
                  gtr: emit(49);
                  geq: emit(50);
                end
              else error(ertyp);
              x. typ := bools
            end
      end (*expression*);
```

```
procedure assignment(lv, ad: integer);
  var x, y: item; f: integer;
  (*tab[i]. obj in [variable, prozedure]*)
begin x. typ := tab[i]. typ; x. ref := tab[i]. ref;
  if tab[i]. normal then f := 0 else f := 1;
  emit2(f, lv, ad);
  if sy = lbrack then
    selector([becomes, eql]+fsys, x);
  if sy = becomes then insymbol else error(erpun);
  expression(fsys, y);
  if x. typ = y. typ then
    if x. typ in stantyps then emit(38) else
    if x. ref<> y. ref then error(ertyp) else
    if x. typ = arrays then emit1(23, atab[x. ref]. size)
  else error(ertyp)
end (*assignment*);

procedure compoundstatement;
begin insymbol;
  statement([semicolon, endsy]+fsys);
  while sy in [semicolon]+statbegsys do
  begin if sy = semicolon then insymbol else error(erpun);
    statement([semicolon, endsy]+fsys)
  end;
  if sy = endsy then insymbol else error(erkey)
end (*compoundstatement*);

procedure ifstatement;
  var x: item; lc₁, lc₂: integer;
begin insymbol;
  expression(fsys+[thensy, dosy], x);
  if not (x. typ in [bools, notyp]) then error(ertyp);
  lc₁ := lc; emit(11); (*jmpc*)
  if sy = thensy then insymbol else error(erkey);
  statement(fsys+[elsesy]);
  if sy = elsesy then
    begin insymbol; lc₂ := lc; emit(10);
      code[lc₁]. y := lc; statement(fsys); code[lc₂]. y := lc
    end
  else code[lc1]. y := lc
end (*ifstatement*);

procedure repeatstatement;
  var x: item; lc₁: integer;
```

```
      begin lc₁ := lc;
        insymbol; statement([semicolon, untilsy]+fsys);
        while sy in [semicolon]+statbegsys do
        begin if sy = semicolon then insymbol else error(erpun);
          statement([semicolon, untilsy]+fsys)
        end;
        if sy = untilsy then
          begin insymbol; expression(fsys, x);
            if not (x. typ in [bools, notyp]) then error(ertyp);
            emit1(11, lc₁)
          end
        else error(erkey)
      end (*repeatstatement*);

      procedure whilestatement;
        var x: item; lc₁, lc₂: integer;
      begin insymbol; lc₁ := lc;
        expression(fsys+[dosy], x);
        if not (x. typ in [bools, notyp]) then error(ertyp);
        lc₂ := lc; emit(11);
        if sy = dosy then insymbol else error(erkey);
        statement(fsys); emit1(10, lc₁); code[lc₂]. y := lc
      end (*whilestatement*);

      procedure forstatement;
        var cvt: types; x: item;
            i, lc₁, lc₂: integer;
      begin insymbol;
        if sy = ident then
          begin i := loc(id); insymbol;
            if i =0 then cvt := ints else
            if tab[i]. obj = variable then
              begin cvt := tab[i]. typ;
                if not tab[i]. normal then error(ertyp) else
                    emit2(0, tab[i]. lev, tab[i]. adr);
                if not (cvt in [notyp, ints, bools, chars]) then error(ertyp)
              end else
              begin error(ertyp); cvt := ints
              end
          end else skip([becomes, tosy, dosy]+fsys, erid);
        if sy = becomes then
          begin insymbol; expression([tosy, dosy]+fsys, x);
            if x. typ <> cvt then error(ertyp);
          end else skip([tosy, dosy]+fsys, erpun);
```

```
    if sy = tosy then
      begin
          insymbol; expression([dosy]+fsys, x);
          if x. typ <> cvt then error(ertyp)
      end else skip([dosy]+fsys, erkey);
      lc₁ := lc; emit(14);
      if sy = dosy then insymbol else error(erkey);
      lc₂ := lc; statement(fsys);
      emit1(15, lc₂); code[lc₁]. y := lc
    end (*forstatement*);

  procedure standproc(n: integer);
        var i, f: integer;
            x, y: item;
    begin
    case n of
1, 2: begin (*read*)
      if sy = lparent then
      begin
        repeat insymbol;
          if sy <> ident then error(erid) else
          begin i := loc(id); insymbol;
            if i <> 0 then
            if tab[i]. obj <> variable then error(ertyp) else
            begin x. typ := tab[i]. typ; x. ref := tab[i]. ref;
              if tab[i]. normal then f := 0 else f := 1;
              emit2(f, tab[i]. lev, tab[i]. adr);
              if sy = lbrack then
                  selector(fsys+[comma, rparent], x);
              if x. typ in [ints, chars, notyp] then
                  emit1(27, ord(x. typ)) else error(ertyp)
            end
          end;
          test([comma, rparent], fsys, erkey)
        until sy <> comma;
        if sy = rparent then insymbol else error(erpun)
      end;
      if n = 2 then emit(62)
    end;
3, 4: begin (*write*)
      if sy = lparent then
      begin
        repeat insymbol;
          if sy = string then
```

```
                              begin emit1(24, sleng); emit1(28, inum); insymbol
                              end else
                           begin expression(fsys+[comma, colon, rparent], x);
                              if not (x. typ in stantyps) then error(ertyp);
                              emit1(29, ord(x. typ))
                           end
                        until sy <> comma;
                        if sy = rparent then insymbol else error(erpun)
                     end;
                     if n = 4 then emit(63)
                  end;
         5, 6: (*wait, signal*)
            if sy <> lparent then error(erpun) else
               begin insymbol; if sy <> ident then error(erid) else
                  begin i := loc(id); insymbol;
                     if i <> 0 then if tab[i]. obj <> variable
                        then error(ertyp)
                        else
                        begin x. typ := tab[i]. typ; x. ref := tab[i]. ref;
                        if tab[i]. normal then f := 0 else f := 1;
                        emit2(f, tab[i]. lev, tab[i]. adr);
                        if sy = lbrack then selector(fsys+[rparent], x);
                        if x. typ = ints then emit(n+1) else error(ertyp)
                        end
                     end;
                     if sy = rparent then insymbol else error(erpun)
                  end;
            end (*case*)
      end (*standproc*);

      begin (*statement*)
         if sy in statbegsys+[ident] then
               case sy of
                  ident: begin i := loc(id); insymbol;
                           if i <> 0 then
                              case tab[i]. obj of
                                 konstant, type1: error(ertyp);
                                 variable: assignment(tab[i]. lev, tab[i]. adr);
                                 prozedure:
                                    if tab[i]. lev <> 0 then call(fsys, i)
                                       else standproc(tab[i]. adr);
                                 funktion:
                                    if tab[i]. ref = display[level] then
                                       assignment(tab[i]. lev+1, 0) else error(ertyp)
                              end
```

```
        end;
    beginsy: if id = #cobegin  # then
        begin emit(4); compoundstatement; emit(5) end
        else compoundstatement;
    ifsy: ifstatement;
    whilesy: whilestatement;
    repeatsy: repeatstatement;
    forsy: forstatement;
    end;
  test(fsys, [ ], erpun)
end (*statement*);

begin (*block*) dx := 5; prt := t;
  if level > lmax then fatal(5);
  test([lparent, colon, semicolon], fsys, erpun);
  enterblock; display[level] := b; prb := b;
  tab[prt]. typ := notyp; tab[prt]. ref := prb;
  if (sy = lparent) and (level > 1) then parameterlist;
  btab[prb]. lastpar := t; btab[prb]. psize := dx;
  if isfun then
    if sy = colon then
    begin insymbol; (*function type*)
      if sy = ident then
      begin x := loc(id); insymbol;
        if x <> 0 then
          if tab[x]. obj <> type1 then error(ertyp) else
            if tab[x]. typ in stantyps then tab[prt]. typ := tab[x]. typ
              else error(ertyp)
      end else skip([semicolon]+fsys, erid)
    end else error(erpun);
  if sy = semicolon then insymbol else error(erpun);
  repeat
    if sy = constsy then constantdeclaration;
    if sy = typesy then typedeclaration;
    if sy = varsy then variabledeclaration;
    btab[prb]. vsize := dx;
    while sy in [proceduresy, functionsy] do procdeclaration;
    test([beginsy], blockbegsys+statbegsys, erkey)
  until sy in statbegsys;
  tab[prt]. adr := lc;
  insymbol; statement([semicolon, endsy]+fsys);
  while sy in [semicolon]+statbegsys do
    begin if sy = semicolon then insymbol else error(erpun);
      statement([semicolon, endsy]+fsys)
    end;
```

```
    if sy = endsy then insymbol else error(erkey);
    test(fsys+[period], [ ], erkey);
end (*block*);

(*---------------------------------------------------------------------- interpret---*)


procedure interpret;
label 97, 98,
const
    stepmax = 8;                    (*max steps before process switch*)
    tru = 1;                                 (*integer value of true*)
    fals = 0;                               (*integer value of false*)
    charl = 0;                          (*lowest character ordinal*)
    charh = 63;                        (*highest character ordinal*)
type ptype = 0. . pmax;                    (*index over processes*)
var ir: order;                               (*instruction buffer*)
    ps:                                        (*processor status*)
    (run, fin, divchk, inxchk, stkchk, linchk, lngchk, redchk, deadlock);
lncnt,                                         (*number of lines*)
chrcnt: integer;                  (*number of characters in line*)
h₁, h₂, h₃, h₄: integer;                        (*local variables*)
s: array[1. . stmax] of integer;                    (*the stack*)

(*process table—one entry for each process*)
    ptab: array[ptype] of record
        t,b,                          (*top, bottom of stack*)
        pc,                                (*program counter*)
        stacksize: integer;                   (*stack limit*)
        display: array[1. . lmax] of integer;
        suspend: integer;             (*0 or index of semaphore*)
        active: boolean                (*procedure active flag*)
    end;
    npr,                      (*number of concurrent processes*)
    curpr: ptype;                       (*current process executing*)
    stepcount: integer;          (*number of steps before switch*)
    seed: integer;                               (*random seed*)
    pflag: boolean;                        (*concurrent call flag*)
procedure setran(seed: integer); extern;
function ran: real; extern;
procedure chooseproc;
(*from a random starting point search for a process that is active and not
    suspended. d aborts the interpreter if a deadlock occurs.*)
var d: integer;
```

```
begin d := pmax+1;
  curpr := (curpr+trunc(ran*pmax)) mod (pmax+1);
  while ((not ptab[curpr]. active)
         or (ptab[curpr]. suspend<>0))
         and (d >= 0) do
  begin d := d−1; curpr := (curpr+1) mod (pmax+1) end;
  if d < 0 then
     begin ps := deadlock; goto 98 end
            else stepcount := trunc(ran * stepmax)
end;


(*functions to convert integers to booleans and conversely*)
function itob(i: integer): boolean;
begin if i=tru then itob: =true else itob: =false end;
function btoi(b: boolean): integer;
begin if b then btoi: =tru else btoi: =fals end;

begin (*interpret*)
  s[1] := 0; s[2] := 0; s[3] := −1; s[4] := btab[1]. last;
  with ptab[0] do begin
  b := 0; suspend := 0 display[1] := 0;
  t := btab[2]. vsize − 1; pc := tab[s[4]]. adr;
  active:=true; stacksize := stmax − pmax*stkincr
                end;
     for curpr :=1 to pmax do with ptab[curpr] do
       begin active := false; display[1] := 0; pc := 0; suspend := 0;
       b := ptab[ curpr−1]. stacksize+1; stacksize := b+stkincr−1;
       t := b−1
       end;
  npr:=0; curpr:=0; pflag:=false;
  seed := clock; setran(seed); stepcount:=0;
  ps := run; lncnt := 0; chrcnt := 0;
  repeat
  if ptab[0]. active then curpr := 0
     else if stepcount = 0 then chooseproc
        else stepcount := stepcount − 1;
  with ptab[curpr] do begin ir := code[pc]; pc := pc + 1 end;
  if pflag then
          begin if ir. f = 18 (*markstack*) then npr := npr + 1;
          curpr := npr
          end;
  with ptab[curpr] do
    case ir. f of
```

```
 0: begin (*load address*) t := t+1;
    if t > stacksize then ps := stkchk
    else s[t] := display[ir.x] + ir. y
    end;
 1: begin (*load value*) t := t+1;
    if t > stacksize then ps := stkchk
    else s[t] := s[display[ir. x] + ir. y]
    end;
 2: begin (*load indirect*) t := t+1;
    if t > stacksize then ps := stkchk
    else s[t] := s[s[display[ir. x] + ir. y]]
    end;
 3: begin (*update display*)
    h₁ := ir. y; h₂ := ir. x; h₃ := b;
    repeat display[h₁] := h₃; h₁ := h₁-1; h₃ := s[h₃+2]
    until h₁ = h₂
    end;
 4: (*cobegin*) pflag := true;
 5: (*coend*) begin pflag := false; ptab[0]. active := false end;
 6: begin (*wait*)
    h₁ := s[t]; t := t - 1;
    if s[h₁] > 0 then s[h₁] := s[h₁] - 1
    else begin suspend := h₁; stepcount := 0 end
    end;
 7: begin (*signal*)
    h₁ := s[t]; t := t-1; h₂ := pmax+1; h₃ := trunc(ran*h₂);
    while (h₂ >= 0) and (ptab[h₃]. suspend <> h₁) do
    begin h₃ := (h₃+1) mod (pmax+1); h₂ := h₂ -1 end;
    if h₂<0 then s[h₁] := s[h₁]+1 else ptab[h₃]. suspend := 0
    end;
 8: case ir. y of
    17: begin t := t+1;
        if t > stacksize then ps := stkchk
          else s[t] := btoi(eof(input))
        end;
    18: begin t := t+1;
        if t > stacksize then ps := stkchk
          else s[t] := btoi(eoln(input))
        end;
    end;
10: pc := ir. y; (*jump*)
11: begin (*conditional jump*)
    if s[t] = fals then pc := ir. y; t := t-1
    end;
```

```
14: begin (*for1up*) h₁ := s[t−1];
      if h₁ <= s[t] then s[s[t−2]] := h₁ else
         begin t := t−3; pc := ir. y
         end
    end;
15: begin (*for2up*) h₂ := s[t−2]; h₁ := s[h₂] + 1;
      if h₁ <= s[t] then
         begin s[h₂] := h₁ ; pc := ir. y end
      else t := t−3;
    end;
18: begin h₁ :=btab[tab[ir. y]. ref]. vsize;
      if t+h₁ > stacksize then ps := stkchk else
         begin t := t+5; s[t−1] := h₁−1; s[t] := ir. y
         end
    end;
19: begin active := true; h₁ := t − ir. y;
      h₂ := s[h₁+4]; (*h₂ points to tab*)
      h₃ := tab[h₂]. lev; display[h₃+1] := h₁;
      h₄ := s[h₁+3] + h₁;
      s[h₁+1] := pc; s[h₁+2] := display[h₃];
      if pflag then s[h₁+3] := ptab[0]. b else s[h₁+3] := b;
      for h₃ := t+1 to h₄ do s[h₃] := 0;
      b := h₁; t := h₄; pc := tab[h₂]. adr
    end;
21: begin (*index*) h₁ := ir. y; (*h₁ points to atab*)
      h₂ := atab[h₁]. low; h₃ := s[t];
      if h₃ < h₂ then ps := inxchk else
      if h₃ > atab[h₁]. high then ps := inxchk else
         begin t := t−1; s[t] := s[t] + (h₃−h₂)*atab[h₁]. elsize
         end
    end;
22: begin (*load block*) h₁ := s[t]; t := t−1;
      h₂ := ir. y + t; if h₂ > stacksize then ps := stkchk else
      while t < h₂ do
         begin t := t+1; s[t] := s[h₁]; h₁ := h₁+1
         end
    end;
23: begin (*copy block*) h₁ := s[t−1];
      h₂ := s[t]; h₃ := h₁ + ir. y;
      while h₁ < h₃ do
         begin s[h₁] := s[h₂]; h₁ := h₁+1; h₂ := h₂+1
         end;
      t := t−2
    end;
```

24: **begin** (*literal*) $t := t+1$;
    if $t >$ *stacksize* **then** $ps := stkchk$ **else** $s[t] := ir. y$
    end;
27: **begin** (*read*)
    if *eof(input)* **then** $ps := redchk$ **else**
      case *ir. y* **of**
        1: $read(s[s[t]])$;
        3: **begin** $read(ch)$; $s[s[t]] := ord(ch)$ **end**;
      end;
    $t := t-1$
    end;
28: **begin** (*write string*)
    $h_1 := s[t]$; $h_2 := ir. y$; $t := t-1$;
    $chrcnt := chrcnt+h_1$; **if** $chrcnt >$ *lineleng* **then** $ps := lngchk$;
    repeat $write(stab[h_2])$; $h_1 := h_1-1$; $h_2 := h_2+1$
    until $h_1 = 0$
    end;
29: **begin** (*write1*)
    if $ir. y = 3$ **then** $h_1 := 1$ **else** $h_1 := 10$;
    $chrcnt := chrcnt+h_1$;
      if $chrcnt >$ *lineleng* **then** $ps := lngchk$ **else**
      case *ir. y* **of**
        1: $write(s[t])$;
        2: $write(itob(s[t]))$;
        3: **if** $(s[t]<charl)$ **or** $(s[t]>charh)$ **then** $ps := inxchk$
          else $write(chr(s[t]))$
      end;
      $t := t-1$
    end;
31: $ps := fin$;
32: $t := b - 1$; $pc := s[b+1]$;
    if $pc <> 0$ **then** $b := s[b+3]$ **else**
      begin $npr := npr -1$; $active := false$;
      $stepcount := 0$; $ptab[0]. active := (npr = 0)$
      end
    end;
33: **begin** (*exit function*)
    $t := b$; $pc := s[b+1]$; $b := s[b+3]$
    end;
34: $s[t] := s[s[t]]$;
35: $s[t] := btoi(\textbf{not}(itob(s[t])))$;
36: $s[t] := -s[t]$;
38: **begin** (*store*) $s[s[t-1]] := s[t]$; $t := t-2$ **end**;

```
45: begin t := t−1; s[t] := btoi(s[t] = s[t+1]) end;
46: begin t := t−1; s[t] := btoi(s[t] <> s[t+1]) end;
47: begin t := t−1; s[t] := btoi(s[t] < s[t+1]) end;
48: begin t := t−1; s[t] := btoi(s[t] <= s[t+1]) end;
49: begin t := t−1; s[t] := btoi(s[t] > s[t+1]) end;
50: begin t := t−1; s[t] := btoi(s[t] >= s[t+1]) end;
51: begin t := t−1; s[t] := btoi(itob(s[t]) or itob(s[t+1]) ) end;
52: begin t := t−1; s[t] := s[t] + s[t+1] end;
53: begin t := t−1; s[t] := s[t] − s[t+1] end;
56: begin t := t−1; s[t] := btoi(itob(s[t]) and itob(s[t+1])) end;
57: begin t := t−1; s[t] := s[t] * s[t+1] end;
58: begin t := t−1;
    if s[t+1] = 0 then ps := divchk else
      s[t] := s[t] div s[t+1]
    end;
59: begin t := t−1;
    if s[t+1] = 0 then ps := divchk else
      s[t] := s[t] mod s[t+1]
    end;
62: if eof(input) then ps := redchk else readln;
63: begin writeln; lncnt := lncnt + 1; chrcnt := 0;
    if lncnt > linelimit then ps := linchk
    end
  end (*case*);
until ps <> run;

98: writeln;
if ps <> fin then
begin
with ptab[curpr] do
  write(#0halt at#, pc:5,# in process#, curpr:4, # because of #);
  case ps of
    deadlock: writeln(#deadlock#);
    divchk: writeln(#division by 0#);
    inxchk: writeln(#invalid index#);
    stkchk: writeln(#storage overflow#);
    linchk: writeln(#too much output#);
    lngchk: writeln(#line too long#);
    redchk: writeln(#reading past end of file#);
  end;
writeln(#0process   active   suspend   pc#);
for h₁ := 0 to pmax do with ptab[h₁] do
  writeln(#0#, h₁: 4, active, suspend, pc);
writeln(#0global variables#);
```

```
      for h₁ := btab[1]. last + 1 to tmax do
        with tab[h₁] do if lev <> 1 then goto 97
        else if obj = variable then if typ in stantyps then
            case typ of
              ints: writeln(name,# = #,s[adr]);
              bools: writeln(name,# = #, itob(s[adr]));
              chars: writeln(name,# = #, chr(s[adr] mod 64));
            end;
      97: writeln
    end (*interpret*);
```

```
(* ----------------------------------------------------------------------- main----*)
```

```
  begin (*main*)
    message(# − concurrent pascal-s#);
      key[ 1] := #and       #;        key[ 2] := #array      #;
      key[ 3] := #begin     #;        key[ 4] := #cobegin    #;
      key[ 5] := #coend     #;        key[ 6] := #const      #;
      key[ 7] := #div       #;        key[ 8] := #do         #;
      key[ 9] := #else      #;        key[10] := #end        #;
      key[11] := #for       #;        key[12] := #function   #;
      key[13] := #if        #;        key[14] := #mod        #;
      key[15] := #not       #;        key[16] := #of         #;
      key[17] := #or        #;        key[18] := #procedure  #;
      key[19] := #program   #;        key[20] := #repeat     #;
      key[21] := #then      #;        key[22] := #to         #;
      key[23] := #type      #;        key[24] := #until      #;
      key[25] := #var       #;        key[26] := #while      #;
      ksy[ 1] := andsy;               ksy[ 2] := arraysy;
      ksy[ 3] := beginsy;             ksy[ 4] := beginsy;
      ksy[ 5] := endsy;               ksy[ 6] := constsy;
      ksy[ 7] := idiv;                ksy[ 8] := dosy;
      ksy[ 9] := elsesy;              ksy[10] := endsy;
      ksy[11] := forsy;               ksy[12] := functionsy;
      ksy[13] := ifsy;                ksy[14] := imod;
      ksy[15] := notsy;               ksy[16] := ofsy;
      ksy[17] := orsy;                ksy[18] := proceduresy;
      ksy[19] := programsy;           ksy[20] := repeatsy;
      ksy[21] := thensy;              ksy[22] := tosy;
      ksy[23] := typesy;              ksy[24] := untilsy;
      ksy[25] := varsy;               ksy[26] := whilesy;
```

```
    sps[# + #] := plus;                    sps[# - #] := minus;
    sps[#(#] := lparent;                   sps[#)#] := rparent;
    sps[# = #] := eql;                     sps[#,#] := comma;
    sps[#[#] := lbrack;                    sps[#]#] := rbrack;
    sps[#"#] := neq;                       sps[#&#] := andsy;
    sps[#;#] := semicolon;                 sps[#*#] := times;
  constbegsys := [plus, minus, intcon, charcon, ident];
  typebegsys := [ident, arraysy];
  blockbegsys := [conststy, typesy, varsy, proceduresy, functionsy,
     beginsy];
  facbegsys := [intcon, charcon, ident, lparent, notsy];
  statbegsys := [beginsy, ifsy, whilesy, repeatsy, forsy];
  stantyps := [notyp, ints, bools, chars];
  lc := 0; ll := 0; cc := 0; ch := # #;
  errpos := 0; errs := [ ]; insymbol;
  t := -1; a := 0; b := 1; sx := 0; c_2 := 0;
  display[0] := 1;
  skipflag := false;
  if sy <> programsy then error(erkey) else
  begin insymbol;
     if sy <> ident then error(erid) else
     begin progname := id; insymbol; end
  end;


  enter(#              #,        variable, notyp, 0); (*sentinel*)
  enter(#false         #,        konstant, bools, 0);
  enter(#true          #,        konstant, bools, 1);
  enter(#char          #,        type1, chars, 1);
  enter(#boolean       #,        type1, bools, 1);
  enter(#integer       #,        type1, ints, 1);
  enter(#eof           #,        funktion, bools, 17);
  enter(#eoln          #,        funktion, bools, 18);
  enter(#read          #,        prozedure, notyp, 1);
  enter(#readln        #,        prozedure, notyp, 2);
  enter(#write         #,        prozedure, notyp, 3);
  enter(#writeln       #,        prozedure, notyp, 4);
  enter(#wait          #,        prozedure, notyp, 5);
  enter(#signal        #,        prozedure, notyp, 6);
  enter(#              #,        prozedure, notyp, 0);
     with btab[1] do
     begin last := t; lastpar := 1; psize := 0; vsize := 0
     end;
```

```
    block(blockbegsys+statbegsys, false, 1);
    if sy <> period then error(erpun);
    if btab[2]. vsize > stmax−stkincr * pmax then error(erln);
    emit(31); (*halt*)
    if not eof(input) then readln;
    if errs = [ ] then interpret else errormsg;
99: writeln
end.
```

BIBLIOGRAPHY

TEXTBOOKS

Concurrent programming is an abstraction of the type of programming that is common in operating systems, and in fact its study is rooted in the difficulties that were encountered in programming operating systems. Thus it is not surprising that other books on this subject deal with operating systems and conversely that books on operating systems usually have a short chapter on concurrent programming. Exceptions are the texts by Holt *et al*. (1978) and Brinch Hansen (1973). The text by Holt describes the construction and use of the CSP/k system for exercising concurrency. CSP/k uses monitors. The text contains a good description of monitor programming and of implementation of concurrency. Its treatment of other synchronization primitives is sketchy.

Brinch Hansen's text is addressed to about the same level as this book. His discussion emphasizes the conditional critical region.

Books on operating systems are those by Tsichritzis and Bernstein (1974) at an elementary level, by Brinch Hansen (1973) and Habermann (1976) at an intermediate level, and by Coffman and Denning (1973) at an advanced level.

One of my favorite books is Brinch Hansen (1977) which contains a description of the language Concurrent Pascal (which uses monitors) and the design and listing of three operating systems.

For an introduction to Pascal, the original reference manual is that by Jensen and Wirth (1975); there are numerous newer texts such as that by Welsh and Elder (1979). An introduction to preliminary Ada is provided by Wegner (1980). There is a newer text by Pyle (1981). For the design and construction of programs see Welsh and McKeag (1980).

SOURCES

The preliminary report on the Ada language is Ichbiah (1979). The manual of the revised language is available from the U.S. Superintendent of Documents, Washington, D.C. Most of the rest of the text is based on the excellent articles by Dijkstra (1968b) and Hoare (1974) both of which should be read. The dining philosophers are discussed in Dijkstra (1971) and Hoare and Perrott (1972).

165

Conway's problem is in Conway (1963) and appears again later in Hoare (1978). While Hoare's system CSP was the basis for the Ada synchronization primitives, CSP is different and worth learning. Most current theoretical work is being done on CSP rather than Ada: Francez (1980) and Apt, Francez and de Roever (1980). Guarded commands were invented by Dijkstra (1975). Another formalism for distributed synchronization has been published by Brinch Hansen (1978).

Lamport's bakery algorithms are described and proved in his papers of 1974, 1977 and 1979. The exercise attributed to Roussel is found in Kowalski (1979). The appendix is based on a program that the Author has used for class exercise for several years. A short description of a preliminary version was published in Ben-Ari (1981). The Pascal-S report has been reprinted in Barron (1981). More sophisticated systems are described in Holt *et al.* (1978) and Kaubisch *et al.* (1976). If you have access to *Concurrent Pascal* (Brinch Hansen, 1977), that is preferable.

The proofs of the programs in the text are semi-formal expositions of formal proofs I have been working on using temporal logic. If this interests you, places to start are Manna and Pnueli (1979) for sequential programs and Pnueli (1981) for concurrent programs. Compare the proof of Dekker's Algorithm given in this book with that in the article by Francez and Pnueli (1978). For another approach to proving concurrent programs see Owicki and Gries (1976).

To test your knowledge of concurrent programming, make the effort needed to understand the mutual exclusion algorithm in Morris (1979).

Aho, A. V., J. E. Hopcroft and J. D. Ullmann, *The Design and Analysis of Computer Algorithms*, Addison–Wesley, Reading, Mass. (1974).

Apt, K., N. Francez and W. P. de Roever, "A proof system for communicating sequential processes", *ACM Trans. on Programming Langauges and Systems*, **2**, 359–385 (1980).

Barron, D. W., *Pascal—The Language and its Implementation*, John Wiley, Chichester (1981).

Ben-Ari, M., "Cheap concurrent programming", *Software—Practice and Experience*, **11**, 1261–1264 (1981).

Brinch Hansen, P., *Operating Systems Principles*, Prentice-Hall, Englewood Cliffs, N.J. (1973).

Brinch Hansen, P., *The Architecture of Concurrent Programs*, Prentice-Hall, Englewood Cliffs, N.J. (1977).

Brinch Hansen, P., "Distributed processes: a concurrent programming concept", *Comm. ACM*, **21**, 934–941 (1978).

Coffman, E. G., Jr. and P. J. Denning. *Operating Systems Theory*, Prentice-Hall, Englewood Cliffs, N.J. (1973).

Conway, M. E., "Design of a separable transition—diagram compiler", *Comm. ACM*, **6**, 396–408 (1963).

Courtois, P. J., F. Heymans and D. L. Parnas, "Concurrent control with readers and writers", *Comm. ACM*, **14**, 667–668 (1971).

Dijkstra, E. W., "The structure of the T.H.E. multiprogramming system", *Comm. ACM*, **11**, 341–346 (1968a).

Dijkstra, E. W., "Cooperating sequential processes" in F. Genuys (ed.) *Programming Languages*, Academic Press, New York (1968b).

Dijkstra, E. W., "Hierarchical ordering of sequential processes", *Acta Informatica*, **1**, 115–138 (1971).

Dijkstra, E. W., "Guarded commands, nondeterminacy and formal derivation of programs," *Comm. ACM*, **18**, 453–457 (1975).

Francez, N., "Distributed termination", *ACM Trans. on Programming Languages and Systems*, **2**, 42–55 (1980).

Francez, N. and A. Pnueli, "A proof method for cyclic programs", *Acta Informatica*, **9**, 133–157 (1978).

Habermann, A. N., *Introduction to Operating System Design*, Science Research Associates, Chicago (1976).

Hoare, C. A. R. and R. H. Perrott (eds.), *Operating Systems Techniques*, Academic Press, New York (1972).

Hoare, C. A. R., "Monitors: an operating system structuring concept", *Comm. ACM*, **17**, 549–557 (1974).

Hoare, C. A. R., "Communicating sequential processes", *Comm. ACM*, **21**, 666–677 (1978).

Holt, R. C., G. S. Graham, E. D. Ladzowska and M. A. Scott, *Structured Concurrent Programming with Operating Systems Applications*, Addison–Wesley, Reading, Mass. (1978).

Howard, J. H., "Proving monitors", *Comm. ACM*, **19**, 273–279 (1976).

Ichbiah J., "Preliminary Ada Reference Manual and Rationale for the Design of the Ada Programming Language", *SIGPLAN Notices*, **14**(6) (1979).

Jensen, K. and N. Wirth, *Pascal User Manual and Report*, Springer-Verlag Lecture Notes in Computer Science, **18**, Berlin (1975).

Kaubisch, W. H., R. H. Perrott and C. A. R. Hoare, "Quasiparallel programming", *Software—Practice and Experience*, **6**, 341–356 (1976).

Kessels, J. L. W. and A. J. Martin, "Two implementations of the conditional critical region using a split binary semaphore", *Information Process. Lett.*, **8**(2), 67–71 (1979).

Kowalski, R., "Algorithm = Logic + Control", *Comm. ACM*, **22**, 424–436 (1979).

Lamport, L., "A new solution of Dijkstra's concurrent programming problem", *Comm. ACM*, **17**, 453–455 (1974).

Lamport, L., "Proving the correctness of multiprocess programs", *IEEE Trans. Software Eng.*, **SE–3**, 125–143 (1977).

Lamport, L., "A new approach to proving the correctness of multiprocess programs", *ACM Trans. on Programming Languages and Systems*, **1**, 84–97 (1979).

Manna, Z. and A. Pnueli, "The modal logic of programs: a temporal approach", *Automata, Languages and Programming, Springer-Verlag Lecture Notes in Computer Science*, **79**, 385–409 (1979).

Morris, J. M., "A starvation-free solution to the mutual exclusion problem", *Information Process. Lett.*, **8**, 76–80 (1979).

Owicki, S. and D. Gries, "Verifying properties of parallel programs: an axiomatic approach", *Comm. ACM*, **19**, 279–285 (1976).

Parnas, D. L., "On a solution to the cigarette smoker's problem (without conditional statements)", *Comm. ACM*, **18**, 181–183 (1975).

Pnueli, A., "The temporal semantics of concurrent programs", *Theoret. Computer Sci.*, **13**, 45–60 (1981).

Pyle, I. C., *The Ada Programming Language*, Prentice-Hall International, London (1981).

Tsichritzis, D. C. and P. A. Bernstein, *Operating Systems*, (*Computer Science and Applied Mathematics Series*), Academic Press, New York (1974).

Wegner, P., *Programming with Ada*, Prentice-Hall, Englewood Cliffs, N.J. (1980).

Welsh, J. and J. Elder, *Introduction to Pascal*, Prentice-Hall International, London (1979).

Welsh, J. and M. McKeag, *Structured System Programming*, Prentice-Hall International, London. (1980).

INDEX